THE ERGONOMICS OF LIGHTING

The Ergonomics of Lighting

by

R. G. HOPKINSON

Professor of Environmental Design and Engineering
University College, London

J. B. COLLINS

Principal Scientific Officer in charge of Lighting Research
Building Research Station
Garston, Watford

MACDONALD TECHNICAL AND SCIENTIFIC
LONDON

SBN 356 02680 9

First published in 1970
Macdonald & Co. (Publishers) Ltd
49–50 Poland Street, London, W.1
Made and printed in Great Britain by
Hazell Watson & Viney Ltd,
Aylesbury, Bucks

Preface

The purpose of this book is to give consultants and environmental designers a summary of studies on the frontiers of lighting technology, particularly those aspects relating the physical environment to the subjective responses of the individual. Designers are now giving much more attention than in the past to the influence of a building on the comfort and working efficiency of the inhabitants.

This emphasis on the relation between the human being and his environment has had a major influence on architecture and planning and has been incorporated into current working technology. At least part of this influence resulted from the cooperation between the research teams at the Building Research Station where the authors were working, and the architectural Development Groups first set up at the Ministry of Education (now the Department of Education and Science), the Ministry of Health and the Nuffield Division for Architectural Studies, and subsequently in other central and local authority organisations. Consequently it is hoped that the book will be of interest not only to lighting technologists, but to a wide range of architects and other consultants to whom lighting is only one of a number of problems which face them.

The daylighting aspects of the lighting problem have already been discussed ('Daylighting,' by R. G. Hopkinson, P. Petherbridge and J. Longmore; Heinemann, 1966), so the present book is confined primarily to the problems associated with artificial lighting, although many of the fundamentals are equally relevant to daylighting.

The academic background of a consultant called in to advise in the design of an environment, especially an industrial environment, can nowadays be equally that of a physiologist, a psychologist, an engineer or an architect. In writing the book, therefore, the authors have thought it wise to allocate some space to a brief review of fundamentals. The first two chapters give a grounding in the rudiments of vision and of lighting engineering. If they are insufficient, there are excellent books to take the reader as far as he wishes to go.

The authors acknowledge with gratitude the contributions which their colleagues made in their investigations. The work on Glare was largely that of Mr P. Petherbridge and Mrs W. Collins, while that on Supplementary Lighting and on Attention was undertaken by Mr J. Longmore. The authors are grateful to Sir Frederick Lea, previously Director of Building Research at the Building Research Station, for his encouragement, and to his successors, for permission to use material first published as communications from the Station.

September 1969

R. G. Hopkinson
J. B. Collins

Acknowledgements

The publishers gratefully acknowledge permission from the following organisations to reproduce the illustrations in this book as follows:

BLACKIE & SON LTD (Fig. 1.3) 'The Perception of Light', by W. D. Wright, 1938.

BRITISH LIGHTING INDUSTRIES LTD (Fig. 10.6).

BRITISH MEDICAL ASSOCIATION (Figs. 3.10, 3.12) 'Measurements of Visual Acuity', by W. R. Stevens and C. A. P. Foxell, from *British Journal of Ophthalmology*, 1955, Vol. 39. Reproduced with the permission of the authors and editor.

ADAM HILGER LTD (Figs. 1.2, 7.1, 7.2) 'The Measurement of Colour', by W. D. Wright, 1964.

ILLUMINATING ENGINEERING SOCIETY, London. (Fig. 3.22) *Trans. Illum. Eng. Soc.* 1961, Vol. 26. (Figs. 3.24, 7.6) *Trans. Illum. Eng. Soc.* 1967, Vol. 32. (Fig. 3.25) *International Lighting Review* 1962, Vol. 13. (Figs. 4.2, 4.3, 4.4) Technical Report No. 10, 1967. (Fig. 8.6) *Trans. Illum. Eng. Soc.* 1958, Vol. 23. (Figs. 9.1, 9.2) *Light and Lighting* 1967, Vol. 60.

ILLUMINATING ENGINEERING SOCIETY, New York. (Figs. 3.15, 3.16, 3.17, 3.18, 3.19, 3.20) *Illuminating Engineering*, June 1959. (Figs. 9.3, 9.4, 9.5, 9.6) *Illuminating Engineering*, May 1966.

H. K. LEWIS & CO. LTD (Fig. 3.6) 'Light, Sight and Work', by H. C. Weston.

MEDICAL RESEARCH COUNCIL (Reproduced with the permission of the Controller of Her Majesty's Stationery Office). (Fig. 3.11) Based on Fig. 9 from Special Report Series No. 173. (Table 7.2) Memorandum No. 43—'Spectral Requirements of Light Sources for Clinical Purposes'. (Fig. 6.1) Report A.P.U. No. 243.

PHILIPS ELECTRICAL LTD (Fig. 7.5) *Philips Technical Review*, 1941, Vol. 6.

POWELL & MOYA (Fig. 10.8).

All the other illustrations are Crown Copyright and reproduced by permission of the Director of Building Research.

Contents

Note on Lighting Units

There is a certain amount of confusion at present over the units which are used for the measurement and calculation of lighting. Part of this is due to the changeover from imperial to metric units and part to the fact that the concept of luminance (brightness) can be expressed either in 'fundamental' units or in terms of equivalent illumination related through reflectance (see Chapters 1 and 2).

In view of the promised change to metric units it has been necessary to use throughout the book a system based on the SI metric system of engineering units, but as this system is not yet in regular use, particularly by engineers and architects, it has also been necessary to express many of the units in terms of the imperial system.

Occasionally it may appear that there is an inconsistency in the conversion factors which have been used between one system and another. The relation between the square foot and the square metre is 10·76 and it has become customary in practical lighting design to use a conversion factor of 10 where great accuracy is not required. Consequently there will be examples in the book where two figures will be given such as '10 lm/ft^2 (100 lux)'. Less frequently some lighting engineers are in the habit of using the rather more accurate conversion figure of 11. This needs to be known since it has been used occasionally in the text.

Again, there are occasions in the text where only the imperial unit is used—this can occur where statutory regulations are quoted, when the regulation gives only the imperial unit value, or when original research results, expressed specifically in imperial units, are quoted.

However, there is a further situation which has arisen through the rewriting of lighting codes in which the new editions are being published in terms of SI units. The drafting committees responsible for drawing up these codes have often taken advantage of the change from imperial to SI units to make alterations to the recommended values themselves, apart from simple conversions. Thus, for example, the IES Code 1968 edition recommends an 'amenity value' of 200 lux which replaces the 'amenity value' of 15 lm/ft^2 in the 1961 edition. This is not an arithmetic conversion but is a new standard. It is, however, sufficiently near to the arithmetic conversion to give rise to possible confusion. Several examples of this kind will be found in the text. These difficulties are inevitable during a changeover period and every effort has been made in the text to avoid confusion or misunderstanding.

Over and above these problems associated with the change from imperial to metric units, however, is the perennial problem of the units to be used for the measurement of luminance. Luminance can be expressed either in terms of luminous intensity per unit area (units, candela per m^2 or ft^2) or in terms of the 'equivalent illumination', expressed in equivalent lux, or apostilbs (asb), or in foot-lamberts (ft-L). Lighting engineers find the 'equivalent illumination' unit essential in their work because it relates the luminance—what the eye sees—to the illumination—the density of physical luminous energy reaching the surface—through the reflectance of the surface, by the simple relation:

$$\text{Luminance} = \text{Illumination} \times \text{Reflectance (numerically)}$$

Thus, a surface of 20% reflectance receiving 100 lux has a luminance of

$$100 \times 0{\cdot}2 = 20 \text{ luminance units (apostilbs)}$$

on the 'equivalent illumination' system.

Unfortunately the only SI unit chosen for luminance is the candela per square metre, and at the time of writing, the equivalent illumination unit (apostilb) is officially deprecated. However, when a similar attempt was made about 50 years ago to eliminate this form of unit, it proved to be so impractical that it was not long before all lighting engineering was again conducted in 'equivalent illumination' units of luminance for obvious reasons of convenience. It is virtually certain that this will happen once again. Therefore it is important that all workers in the field of lighting and vision reading this book should keep in mind the simple relationship between the two units:

$$1 \text{ apostilb} = 1/\pi \text{ cd/m}^2 = 0{\cdot}3183 \text{ cd/m}^2$$

It will be appreciated from the above that the situation regarding lighting units is at present hopelessly confused except to those who have grown up with the various changes which have taken place in the system during the present century, and the situation has not been helped by the well-meaning activities of those responsible for the present SI unit of luminance. In order to assist the reader in finding his way through this confused situation a table has been drawn up which may help where the text itself still leaves the matter in some doubt.

Conversion Table

Illumination

1 lm/ft^2 = 10·76 lux
1 lux = 0·0929 lm/ft^2
The conversion 1 lm/ft^2 = 10 lux is often used.

Luminance

1 candela/ft^2 = 10·76 candela/m^2
1 candela/m^2 = 0·0929 candela/ft^2
1 candela/ft^2 = π foot-lamberts
1 candela/m^2 = π apostilbs

No. of → multiplied by the factor gives the no. of ↓	*Foot-lambert (ft-L) or equivalent foot-candle (e.f.c.)*	*Apostilb (asb) or equivalent lux*	*Candela per m^2 (cd/m^2) or nit (nt)*	*Candela per ft^2 (cd/ft^2)*	*Candela per in^2 (cd/in^2)*
ft-L or e.f.c.	1	0·0929	0·2919	3·142	452
asb	10·76	1	3·142	33·82	4870
cd/m^2 (nit)	3·426	0·3183	1	10·76	1550
cd/ft^2	0·3183	0·02957	0·0929	1	144
cd/in^2	0·00221	0·0002	0·00065	0·0069	1

It is customary in rough calculations to round the above conversion factors to the nearest 1 or 3. Thus

$$1 \text{ cd/m}^2 = \tfrac{1}{3} \text{ ft-L} = 3 \text{ asb} = 0{\cdot}1 \text{ cd/ft}^2 \quad \text{approx.}$$
$$1 \text{ ft-L} = 10 \text{ asb} = 3 \text{ cd/m}^2 = 0{\cdot}3 \text{ cd/ft}^2 \quad \text{approx.}$$

PART I. INTRODUCTION

1

Visual Functions in Relation to Lighting

The science and art of lighting depend upon the adaptability of the eye to function with reasonable efficiency in a wide range of conditions. The eye is able to make use of very high and of very low levels of light, but there are certain optimum conditions in which it works best, and the aim of lighting research is to find out exactly what are these optimum conditions. The aim of good lighting is to provide them.

No theory of lighting can succeed which looks upon the eye merely as a camera. The eye is, indeed, an optical instrument which performs very much like a camera, and the analogy is a useful one for understanding certain aspects of vision. On the other hand, the eye itself is merely the organ which transmits visual information to the brain to interpret in terms of past experi-

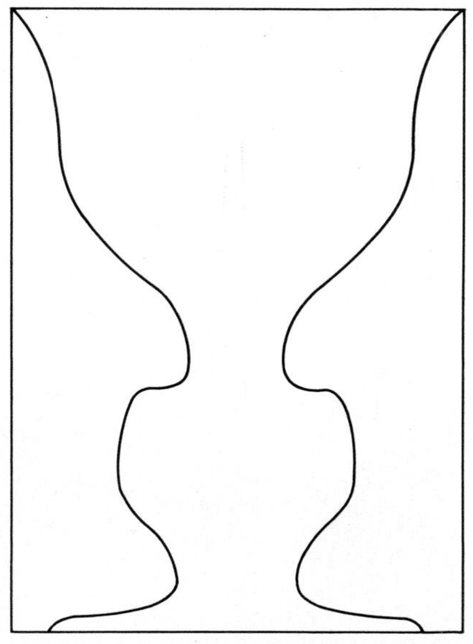

Fig. 1.1 *Ambiguous information. The eye sees what the brain chooses to see. A camera has no choice*

ence. The information contained in Fig. 1.1 depends upon the interpretation which the brain puts upon the ambiguous information; what is seen first, the two faces or the urn, is not a function of the acuteness of vision nor of the lighting conditions. A camera would record the information as it is presented in Fig. 1.1. It is the brain behind the eye which decides what shall be seen and what shall be rejected.

The optical system of the eye, which determines the function of the eye as a camera, is relatively well understood. The function of the brain as a visual computer is only beginning to be barely sensed in a dim, elementary and greatly wondering manner. In writing of the eye as a camera, one has the backing of much experimental fact. To write of the eye as a computer demands much speculation.

Fig. 1.2 is a simple diagram of the structure of the eye. The outer layer consists of the sclera, a hard structure which is modified at the front of the eye into a transparent layer, the cornea, the whole protecting the inner eye from damage. Immediately under the sclera is the choroid, a black membrane which absorbs light and prevents inter-reflections, and which is modified at

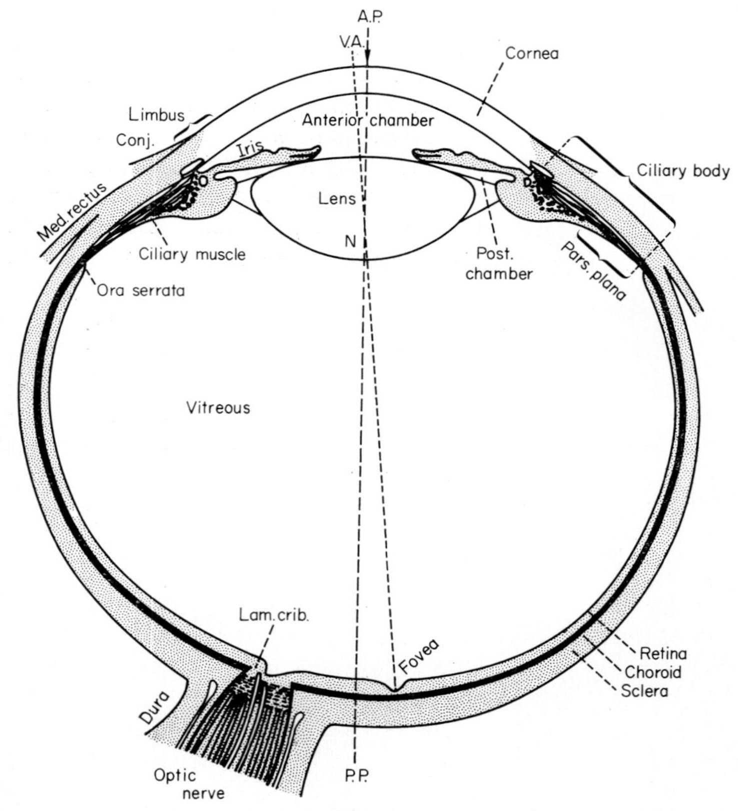

Fig. 1.2 *A horizontal section of the human eye (From Wolff's 'Anatomy of the Eye and Orbit')*

the front of the eye into the transparent lens, the purpose of which is to focus light accurately on to the third layer, the retina, where are located the sensitive light receptors of the eye. The focal length of the lens is adjusted by means of the ciliary muscle which changes the curvature of the lens and so produces a sharp image either for objects close to, or distant from the eye. This *accommodation* process is automatic in the healthy eye. Between the lens and the cornea is the iris or pupil which is controlled by the iridomotor muscles, the normal function being to reduce the diameter of the pupil in bright light in order to protect the retina. The pupil diameter also changes when the eye accommodates for near objects, and this has the advantage of 'stopping down' the lens and so giving a higher degree of resolution.

The interior of the eyeball is filled with a fluid which is liquid in the chamber in front of the lens (the 'aqueous humour'), and jelly-like between the lens and the retina (the 'vitreous body'). Light has to pass through this fluid on its way to the retina, and according to its clarity (which becomes less in old age) a small amount of light is scattered out of the direct path. It is this scattered light which reduces the contrast in the dark parts of a scene with a large range of brightness.

The retina, which in its construction is an extension of the brain itself, contains a very large number of very small light receptors called, because of their shape, the *rods* and the *cones*, which each have a special function to perform. The cones, which are distributed chiefly around the centre of the eye and which become more closely packed towards the centre, perform the function of precise seeing at normal brightness levels. The cones are also the seat of colour perception. The rods, on the other hand, are distributed more widely around the periphery of the eye; they are not colour-sensitive but have an extremely high sensitivity, so that their function is to permit seeing at very low levels of light. The rods are far less closely packed than the cones, and so the resolution of detail by rod vision is very much less precise than with cone vision. In other words, the eye sees best centrally in bright light, its resolution is very high and it sees colour, whereas in the near-dark, resolution is poor, colour vision is absent, and vision, though nowhere good, is better away from the centre. Anyone who has much seeing to do in the dark knows that he sees better if he looks slightly away from the object of regard.

The cones are connected to the visual centres of the brain by a direct one-to-one link. The rods, on the other hand, are linked in groups to a single nerve path, so that these two factors taken together with the closer packing of the cones exaggerate even more the disparity between the resolution of the eye by cone vision and by rod vision. The arrangement of the rod receptors has a special purpose. Because many rods are linked to one nerve path, there is a high chance that an image falling upon the periphery of the eye will register a sensation. When man depended upon his visual alertness for his survival, it was a great advantage to him when going abroad at night to be warned of danger in the form of a moving predator. On the other hand, during the daytime his need was for himself to see his own prey at the greatest possible distance, and so his highly acute central cone vision served its purpose.

In the industrialised world relatively little use is made of pure rod vision. Indeed many of its more remarkable characteristics were 'rediscovered' during the blackout of the Second World War, and many experts on vision found themselves having to study in renewed detail attributes of vision which had ceased to be common knowledge to the urban dweller. Even when driving a car on the streets at night, pure rod vision barely comes into operation. The central visual acuity given by the rods is inadequate for safe driving (less than one tenth of daytime normal, and less than one quarter of the minimum required by law in Great Britain for the holder of a driving licence). Visual acuity at night must be aided either by the use of headlamps on the car itself, or by the provision of roadway lighting. Both of these artificial aids raise the luminance on the carriageway well above the pure rod vision level.

Adaptation and Luminance Range

Adaptation to the light or to the dark does not take place instantaneously. The adaptation process is complex and is believed to take place in several stages. One of these stages involves photochemical transformations in the retina, and some of these transformations have time constants measured in minutes, while others take place more quickly. If one goes into a darkened

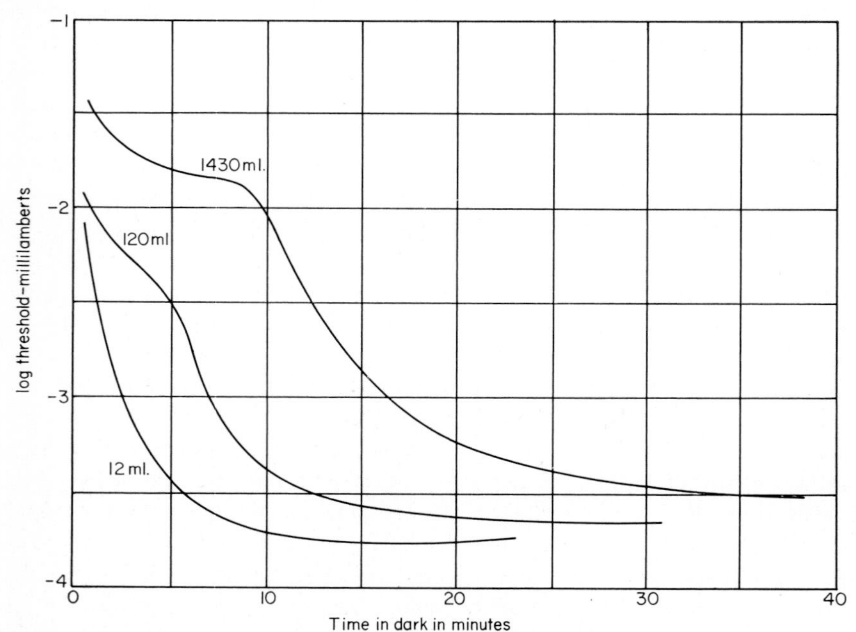

Fig. 1.3 *Variation of dark-adaptation with intensity of previous stimulation (Winsor and Clark 1936). Initial intensity of light adaptation shown in diagram*

1 millilambert is approximately equal to 1 foot-lambert

room where the level of lighting is comparable with starlight, the eyes will take from ten to thirty minutes to adapt fully and to allow the maximum visual information to be obtained from the very low level of lighting. The course of dark adaptation can be determined as a function of the contrast threshold, that is, the least contrast above the adaptation level which can be detected. The progress of adaptation as determined in this way follows a curve of the type shown in Fig. 1.3. The first stage of the curve represents the dark adaptation of the cone system of the retina, which may take of the order of one to five minutes. The discontinuity in the dark adaptation curve represents the point at which the rod mechanism comes into operation. For the rods to be fully dark-adapted may take another ten to thirty minutes. Subjectively this phenomenon is well known. One goes into a darkened room, say a cinema or auditorium, and for a few minutes one feels completely blind. Then almost suddenly, objects around become visible. This is the point where the rod mechanism takes over and corresponds to the sharp drop in the dark adaptation curve of Fig. 1.3.

Vision in the light by means of the cone mechanism is called 'photopic vision' while vision in the dark by means of the rods is called 'scotopic vision'. Moonlight can be taken as the upper limit of scotopic vision. The test is whether colours can be recognised or not. There is a region between moonlight up to twilight where neither the rods nor the cones are fully operative. Twilight vision is called 'mesopic vision' and its characteristics both as regards adaptation, colour perception, visual acuity, and contrast sensitivity are very

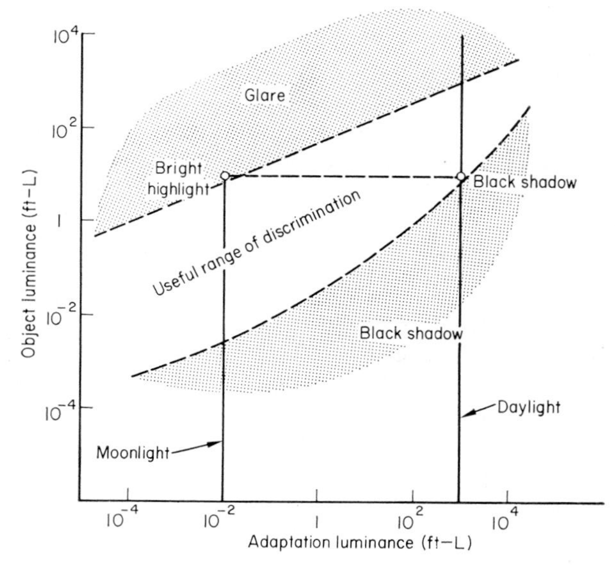

Fig. 1.4 *Schematic diagram showing the range of discriminable luminance at any given level of adaptation*

The limit lines shown are in no sense sharp boundaries—glare and loss of highlight detail gradually increase as luminance increases; loss of shadow detail gradually merges into subjective black as luminance decreases

difficult to define in precise terms, but lie between the true scotopic and true photopic characteristics (Palmer 1966).

The range of luminance which the eye can perceive extends, therefore, from shadows in starlight to the brightness of the snow in full sunlight. This range in physical terms is enormous, of the order of 10^{12} to 1. The eye cannot, however, perceive this full range at any one instant. Instead it operates like an instrument which has a built-in sensitivity which automatically adjusts itself to the average conditions. In daylight the eye adapts to the average of the scene around it, and it can then perceive a luminance range of about 100 to 1 in bright sunlight, or perhaps 1000 to 1 in dull daylight. Fig. 1.4 shows that, for example, a physical luminance at the lower end of this daylight range, which appears as a black shadow to the daylight-adapted eye, will appear as a bright highlight to the eye adapted to moonlight. Physical luminance therefore correlates only vaguely with the subjective sensation of brightness. In order to explain in quantitative terms what the eye actually sees when it receives light from a surface, it is necessary to introduce the concept of *apparent brightness*, that is, the quantitative expression of luminous sensation in terms both of the physical luminance of the object of regard and the physical luminance of the adaptation level of the eye at the time. The relation between physical luminance, adaptation luminance, and apparent brightness is dealt with in greater detail in Chapter 10.

Light and Sight

Sight is deeply dependent upon lighting level, both as regards the ability to see fine detail and the ability to appreciate brightness differences or contrasts. The relation between these factors is discussed at length in Chapter 3, and so it is sufficient to record here that both contrast threshold sensitivity, that is, the ability of the eye to detect differences in luminance such as a piece of white thread against a near-white piece of cloth, and also sharpness of vision (visual acuity), improve progressively as the luminance is raised from low levels to high levels. The eye can detect detail at least ten times as fine in daylight as it can in moonlight. Its ability to detect contrasts also varies in about the same ratio. Since the eye also has colour vision which it does not have in moonlight, the total visual capacity of the eye in daylight is obviously very much greater. The dependence of the seeing power of the eye upon lighting level is now the basis of most codes of recommended lighting practice throughout the world (see Chapter 9).

The eye not only sees contrast and detail better in good light. Moving objects can be seen with more certainty and so necessary action can be anticipated in good time. A given object is not only seen more quickly, but significant information is picked up more quickly from a complicated seeing task when the lighting is good than when it is poor. This, of course, is well known. In poor lighting one can read fine print painstakingly and with effort, and possibly making some errors, thus undertaking slowly a task which, in full daylight, could be done in a small fraction of the time. Ball games which call for quick anticipation cannot be played except in good lighting. Bad light stops play.

Vision in the Elderly

Young children have the best vision in every respect. Their maximum acuity is greater, their range of adaptation is greater, they can accommodate to objects very close to their eyes and they can also see clearly at a distance, while their optic media are clear so that they are not troubled unduly by scattered light.

It does not follow, however, that young children will show up better than adults in any of the normal tests of vision. The adult has greater experience, he can make intelligent guesses at detail which he may not be able to see precisely, and so his combination of a slightly ageing optical instrument with a very much more experienced and efficient brain computer may result in an overall visual capacity greater than that of a child. Nevertheless in certain respects he is unable to compensate for changes which take place in the older eye.

The most important of the deficiencies related to age is the inability to see clearly objects close to the eye. From childhood the 'near point' gradually recedes until, for normal vision, it has receded by the age of 50 to a point well beyond the comfortable reading distance. A person of this age often complains that although his sight is as good as ever, his arms are no longer long enough. In addition to this recession of the near point, certain changes take place in the retina of the eye which result in the eye requiring more light in order to be able to see as well. For the same visual performance, therefore, the older eye requires more light to supplement the essential reading spectacles.

Certain changes also take place in the eye which result in more light being scattered in the optic media, and so the elderly person is more affected by stray light or a glaring light in his field of view. His colour vision is also slightly modified by a yellowing of the lens and of central portions of the retina, both of which result in a decreased sensitivity in the blue region of the spectrum.

Lighting and Defective Vision

It is only rarely that the lighting designer has to consider special lighting for people who suffer from serious visual defects. When this is necessary, as for example, in the lighting of schools for partially-sighted children, or in the lighting of geriatric wards in hospitals, specialist advice is clearly necessary.

Very few visual defects of a serious nature can be helped by modifications in a normal scheme of lighting. Some forms of visual defect can, however, be assisted by the special use of light, and in the case of a partially-sighted person, this may mean the difference between living in a visual and a non-visual world. Some people with gross visual defects can be enabled to read by a combination of carefully screened local lighting on their work and the use of special reading material; for example, many elderly cataract patients can see to read white letters on a black ground if they have sufficient light, which is concentrated on the book and which cannot enter their eyes. But it is still true that the use of light to aid poor sight needs special care and consideration for every individual case, because what may aid one may be useless for another.

The common refractive disorders of vision, short sight, long sight, and astigmatism, can be easily corrected by suitable spectacles so that the patient then has, to all intents and purposes, normal vision. Reading spectacles for the elderly also permit normal or near-normal vision to be attained for near work.

Non-refractive disorders such as cataract and nystagmus can be assisted to some slight degree by the provision of special lighting concentrated on the work. Sufferers from very high myopia (short sight), whose vision cannot be corrected fully by spectacles, also reap some benefit from selective lighting to high levels.

Binocular defects of vision cannot, however, be materially assisted by lighting. The two eyes normally work together so that they can *converge* on near objects, or the convergence can be relaxed to distant objects without the observer being conscious of any kind of double vision. Where the oculomotor muscles are perfectly adjusted, such convergence changes can be made with no effort and no resultant fatigue. Where there is a muscular imbalance, however, double vision may result. This may cause the vision of one eye to deteriorate and to be suppressed, leaving the other to do all the work. Where the imbalance is slight, binocular vision may be retained but the effort required to preserve it may be very great and may result in ocular muscle fatigue, headache and other symptoms. Unfortunately lighting can do little to alleviate such a condition.

Equally, lighting can do nothing to assist defects in colour vision. So-called 'colour blindness' is not blindness as generally understood, but it results from the cone mechanism, which governs colour vision, being defective in some way. As a result, colours which are easily distinguishable by normal sight are confused by the colour blind. The defect is caused by the absence of one or more of the colour sensation mechanisms present in cone vision, and so there is nothing that can be done, either by the provision of coloured spectacles, or by the use of specially coloured illuminants to restore colour vision to normal. Curiously, some colour blind people can see through some forms of camouflage which deceive the normal eye, but otherwise the colour blind will always be at a disadvantage to the normal, except in monochromatic lighting like that given by the sodium discharge lamp, where the normal eye has no sense of colour and so is for once at no advantage over the colour blind eye.

Flicker and Intermittent Light

A light source operating on alternating current will vary in its light output in phase with the frequency of the AC supply. The eye has the power of integrating a regularly pulsating luminance in such a way that if the frequency of pulsation is beyond the 'critical fusion frequency', the light will be seen as steady. This critical fusion frequency depends upon a number of factors, particularly the luminance, the size of the source of intermittent light, and the wave form of the light output with respect to time. If the frequency of alternation is below the critical fusion frequency (CFF) the light will be seen to flicker. Flicker is more readily detectable in the periphery of the eye than in the central field.

Flicker sensitivity also varies considerably from person to person, and it also varies in the same individual depending on his state of health or state of fatigue. Flicker sensitivity has, in fact, been used as a means of assessing both general and visual fatigue (see Chapter 6).

Flicker can be a nuisance to some sensitive people, and a severe disability to others, and the remedy is for them to avoid working in lighting which troubles them, and of course, not to watch television or the cinema. The alleviations of the flicker problem are discussed in more detail in Chapter 5.

Glare

The eye does not function at its full efficiency when there is unwanted light in the visual field. A bright window or a lighting fitting can cause a situation of unfavourable adaptation, which may be described by the sufferer as 'uncomfortable' or 'blinding' or 'dazzling', and the words he uses suggest that his vision is impaired, or that he is suffering discomfort or pain.

When there is unwanted light in the eye, one result may be that this light may be scattered in the optic media, especially in elderly eyes, and this causes a veiling brightness which interferes with clear vision. Alternatively, the unwanted light may affect the level of adaptation and the eye may adjust its sensitivity to the bright light and so lose its sensitivity to the shadows in the scene. In either case the result will be the same, in that vision will be affected in the shadows and the less well lighted parts of the scene. When this happens a state of 'disability glare' is said to exist. Oncoming headlights on a dark night in an unlighted street cause serious disability glare.

Immediately after bright lights are removed, there may be a residual reduction in vision which persists until the adaptation has recovered. This effect is sometimes called 'successive glare' and is an adaptive effect. Whereas the 'veiling' effect of stray light in the eye disappears immediately with the cause, the successive effect takes a little time for the photochemical processes in the eye to adjust to a new balance following the removal of the original cause of glare.

A form of glare can arise without any accompanying disability, however;

in a well-lit interior the light sources can cause discomfort without causing any detectable diminution in the ability to see. This situation is called 'discomfort glare' but its origin is not known. It is not due to scattered light, and it is not a successive effect because immediately the cause is removed, the discomfort ceases. It appears to arise in two different types of situation, although both are often present together; one of these occurs when the whole visual field is so bright that to look anywhere causes discomfort—sunshine on a landscape covered in white snow is the classic case; the other is the contrast effect which arises when there is a very bright area in an otherwise much darker scene—bright clouds seen through a window in a rather dark room is an example. The first effect, the 'saturation' effect, is believed to arise because the field is bright beyond the limit within which the retinal elements function efficiently, with the result that the brain sends instructions to the iridomotor muscles still further to reduce the pupillary diameter to reduce the retinal illumination; if the iris is already at its least diameter, a state of muscular imbalance is set up (the pupil can be seen to oscillate in diameter) and it is believed that this is at least partly responsible for the discomfort. With the second effect the greater part of the visual field is well within the adaptation range of the retinal mechanism, but the glare source is beyond the range to which the eye is adapted at the time. It is believed that in this case also the iridomotor muscles also reach a state of imbalance, since in severe cases of 'contrast discomfort' the pupil can also be observed in a state of oscillation, although no effect can be detected in moderate degrees of discomfort.

Both forms of discomfort are of great importance in the design of good lighting, and they will be discussed in Chapter 4.

Fixation, Attention and Distraction

There is a distinction between 'seeing' and 'looking'. In order to 'look' the eye and brain together make a decision to concentrate the attention upon one part of the whole visual field and to select for the moment this area for the gathering of visual information. This act of 'fixation' is usually a deliberate act of will; the eye, and head if necessary, are turned in the direction of the wanted area, which once found is retained on the central part of the retina where vision is best. In performing this deliberate act, the eye makes scanning movements—'saccades'—which are designed to permit the searching to be done most efficiently. The act of fixation is an important visual function. It is normally done without difficulty or effort.

The eye can, however, be distracted from its acts of voluntary fixation by areas of high brightness or contrast or colour, or any combination of the three. High brightness is the greatest distraction. This is because there remains in the eye the effect known in plants and animals as 'phototropism', the involuntary turning towards the light. Oncoming headlights on a dark night are an almost insuperable phototropic distraction. It also follows that the attention will be more easily held by a very bright area than by a dark area—it is more difficult to fix the attention on a shadowed area than a highlight. The significance of these visual phenomena to lighting are discussed later.

References

Palmer, D. A. (1960): 'A System of Mesopic Photometry'. *Nature*, **209**, 276–281.
Winsor, C. P. and A. B. Clark (1936): 'Dark Adaptation after Varying Degrees of Light Adaptation'. *Proc. Nat. Acad. Sci.*, **22**, 400.
Wright, W. D. (1964): 'The Measurement of Colour'. Adam Hilger, London.

Further Reading

Gregory, R. L. (1966): 'The Eye and Brain'. World University Library, London.
Hopkinson, R. G. (1963): 'Architectural Physics: Lighting'. HMSO, London.
Weale, R. A. (1963): 'The Aging Eye'. H. K. Lewis, London.
Weston, H. C. (1962): 'Light, Sight and Work'. H. K. Lewis, London.
Wright, W. D. (1938):'The Perception of Light'. Blackie, Glasgow.

2

Rudiments of Lighting Practice

Light and Radiation

The band of electromagnetic radiation which lies in the wavelength band from approximately 400 nm* to 750 nm gives rise to a sensation of light in the human eye. The eye is not equally sensitive to all radiation within this band, but has a peak sensitivity at about 555 nm. The relative spectral luminosity of the light-adapted eye (photopic vision—see Chapter 1) is as shown in Fig. 2.1 (curve A).

The sensation of colour is associated with the wavelength of radiation. The approximate sensation given by radiation in different spectral bands is as shown in Fig. 2.1. Neither ultraviolet radiation, of wavelength shorter than 400 nm, nor infrared radiation, of wavelength longer than 750 nm, causes any sensation of light in the normal eye.

Under very feeble conditions of lighting, such as moonlight or starlight, the eye is preferentially more sensitive to radiation of shorter wavelength. The relative spectral luminosity of the dark-adapted eye (scotopic vision—see Chapter 1) is as shown in curve B of Fig. 2.1, where it may be compared with curve A. In the dark-adapted condition the eye has no sense of colour.

The luminous efficiency of a source of light is governed by the amount of radiation which it emits in the visible range of the electromagnetic spectrum relative to the radiation which it emits at all wavelengths. This is usually related to the amount of power consumed when it is emitting, and the luminous efficiency of a source is usually expressed in terms of the number of units of light emitted per unit of power. Lighting engineers, preferring to regard efficiency purely as a ratio (with a maximum of 100%) now use the term 'efficacy' for this quantity.

The unit of light is the *lumen*. The luminous efficacy of a light source such as an electric lamp is therefore expressed in terms of lumens per watt (lm/W).

The two forms of artificial light source in general use today are the incandescent filament lamp in its various types, and the gaseous vapour discharge lamp, either without a fluorescent envelope as in the sodium vapour discharge street lighting lamp, and the high-pressure mercury vapour discharge street lighting lamp, or with a fluorescent envelope coated with a

* 1 nanometre or nm = 10^{-9} metre.

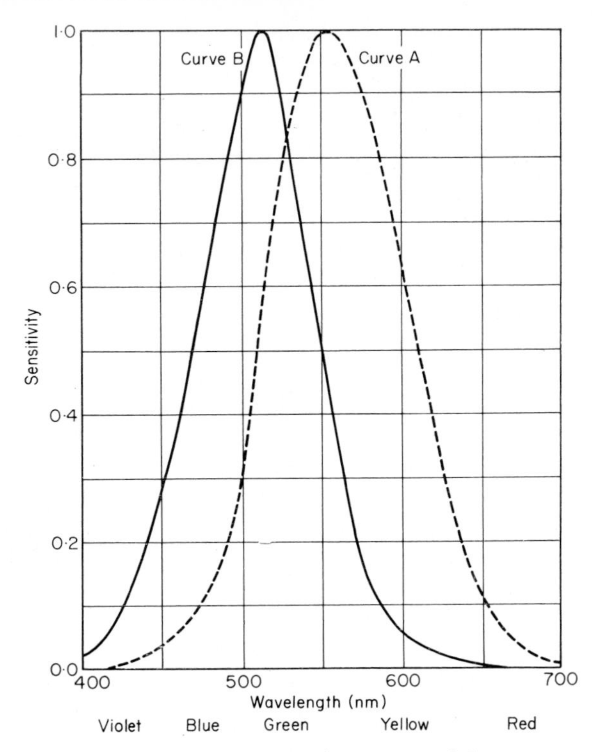

Fig. 2.1 *Relative spectral sensitivity of the eye*
Curve A: Light-adapted
Curve B: Dark-adapted

material which converts the invisible ultraviolet radiation of the mercury vapour discharge into visible radiation of the required spectral distribution.

The physics of the incandescent lamp is based on Planck's quantum theory of radiation by which the spectral energy distribution of radiation from a theoretically perfect radiator can be predicted. Such a theoretically perfect radiator is called a 'black-body radiator'. An incandescent lamp has been shown experimentally to approximate sufficiently closely to a perfect radiator for its light emitting properties to be predicted with adequate accuracy from considerations of black-body radiation.

The total radiant power per unit area of a black-body radiator varies as the fourth power of the absolute temperature. The spectral emitting characteristics of an incandescent lamp can thus be largely defined by specifying the 'correlated colour temperature' (CCT) of the lamp, that is, the temperature at which a perfect radiator would be operated to give the closest possible colour match to the light from the incandescent lamp (see Chapter 7).

The fluorescent lamp is basically a discharge in ionised mercury vapour between two electrodes located at either end of a long tube; the discharge generates ultraviolet radiation which in turn is converted by the fluorescent phosphor coating of the tube into visible light. The colour of the light and the luminous efficacy of the lamp depend upon the spectral emission of the fluorescent material. Provided the total radiation from the lamp (from the mercury discharge itself and from the fluorescent coating taken together) covers a sufficiently wide range of the visible spectrum, the resultant colour will be sufficiently close to that of a black-body radiator for a crude specification of the colour appearance of the lamp to be given in terms of its correlated colour temperature.

The efficiency of energy conversion of both an incandescent lamp and a fluorescent lamp is low because a great deal of the radiation from the lamp is in the form of non-visible radiation, particularly heat. The theoretical maximum possible luminous efficacy for a light source is 680 lm/W and this value would only be obtained if all the radiation were emitted as yellow-green light at 555 nm wavelength, that is, at the wavelength of maximum spectral sensitivity of the human eye. The maximum possible efficacy of any white light source, with its energy distributed throughout the visible band of the radiation spectrum only, is approximately 200 lm/W.

The way in which practical light sources so far produced fall short of the best possible performance can be seen by comparing these figures with luminous efficacies of lamps commercially available. For a commercial 'white' fluorescent lamp, this may be 60 lm/W, or for a high-pressure sodium discharge street lighting lamp (the type with the golden-white colour which is succeeding the monochromatic yellow low-pressure type) 100 lm/W. The ordinary general service type of incandescent lamp, on the other hand, may have an efficacy of the order of only 10 lm/W, although specialised tungsten-halogen ('quartz-iodine') types can have a much higher efficacy.

The luminous efficacy of an incandescent filament lamp is closely associated with the operating conditions, since the luminous output varies as the fourth power of the absolute temperature of the filament. If the lamp is run at low power at a low colour temperature, it will radiate proportionally more heat than light and its luminous efficacy will be low. If the same lamp is run at higher power and so at a higher temperature, its luminous efficacy will be higher. If run at too high a temperature, the tungsten filament of the lamp will vaporise and the lamp will fail. The operation of an incandescent filament lamp is a compromise between low luminous efficacy and catastrophically short life. Most filament lamps for general service are run with a power input to permit them to emit at a colour temperature of about 2600K, giving them a luminous efficacy in the region of 10–12 lm/W, and an operating life of about 1000 hours. The average light output of a lamp over its useful life for design purposes is stated in the relevant British Standards—161 (1968) and 1853 (1967)—as 'Lighting Design Lumens'.

The luminous efficacy of a fluorescent lamp is much less dependent upon its operating conditions, but depends greatly upon the nature of the fluorescent coating. The design of a fluorescent lamp is a compromise between high luminous efficacy and the accurate rendering of colours. With the luminous materials at present available for coating the lamp, it is possible to design a lamp which matches daylight tolerably well in colour appearance and rendition of surface colours while having a luminous efficacy of the order of 40 lm/W. By sacrificing accuracy of colour rendition, lamps can be obtained which look tolerably white while having a luminous efficacy of the order of 60 lm/W.

The operating life of a fluorescent lamp is not related to the correlated colour temperature of the light emitted, but is a function of the design of the electrodes and of other characteristics. A good fluorescent lamp has a useful life of the order of 10 000 hours and does not fail catastrophically at the end of this period, but continues to emit less and less light until it is no longer economical to continue with its use. The life is much reduced by frequent switching, however.

The light output of a fluorescent lamp is more dependent upon the ambient temperature than that of an incandescent lamp. The lamp should be operated at the ambient temperature which results in the correct pressure of the mercury vapour in the lamp. This temperature is usually of the order 25–30°C. Temperatures much lower or much higher than this result in a marked reduction in luminous efficiency.

Basic Photometry Units

The standard of light at present in use is defined in terms of the light emitted by a black-body radiator at the freezing point of platinum (2040K). Such an emitter is defined as emitting 60 *candelas* per cm^2 of its luminous surface area.

The *lumen* is defined as the luminous flux emitted through one steradian from a uniform point source of light of intensity 1 candela.

The *luminous intensity*, expressed in candelas, is conversely the luminous flux, expressed in lumens, per unit solid angle in a given direction.

The *illumination* is the spread of light over a surface, and is therefore expressed in lumens per unit area. In the metric system, the unit is the *lux*, which is one lumen per square metre (lm/m^2) while in British units the unit is the lumen per square foot (lm/ft^2). Since 1 m^2 = 10·76 ft^2, the two units are related by:

$$1 \text{ lm/ft}^2 = 10{\cdot}76 \text{ lux}$$

(It is useful for rough estimation to take 1 lm/ft^2 as 10, or more nearly 11, lux.) Thus a point source of luminous intensity of 1 candela, radiating light flux in all directions, placed at the centre of a sphere of unit radius (say 1 metre) will illuminate the surface of the sphere to a level of 1 lux (i.e. 1 lm/m^2). Since the sphere has a total area of 4π m^2 (12·57 m^2), the source of intensity 1 candela emits 12·57 lumens.

Luminance is the physical (photometric) correlate of the psychological sensation of *brightness*. A theoretically perfect diffusing white surface receiving an illumination of 1 lm/ft^2 has a luminance of 1 *foot-lambert* (ft-L); if it receives 1 lux, it has a luminance of 1 *apostilb* (asb).

Thus 1 ft-L = 10·76 asb or 10 asb roughly.

The Two Laws of Illumination

1. *The inverse square law of radiation* states that the illumination E at a point on a surface varies inversely as the square of the distance d between the surface and the point source of light which illuminates it. A surface normal to the direction of the incident light has an illumination expressed by:

$$E = I/d^2 \text{ lumens/unit area,}$$

where I is the intensity in candelas of the point light source.

2. *The cosine law of illumination* states that the illumination on any surface varies as the cosine of the angle of incidence of the surface to the direction of light. Thus if the angle between the normal to the surface and the direction of the incident light is θ, the illumination is given by:

$$E = I \cos \theta / d^2 \text{ lumens/unit area}$$

The Calculation of Illumination

The level of illumination expressed in lumens per unit area can be deduced from the inverse square law and the cosine law as indicated above. The illumination on a working surface from a single small source of light such as a filament lamp would be calculated in this way.

The precision to which lighting calculations are made must always be defined. In the limit, such calculations are concerned with visual sensations of light, and the eye is not sensitive to small differences in the amount of light. Practical lighting calculations take this into account; unnecessary precision is a waste of effort. On the other hand, light has to be paid for, and careless calculation can waste useful resources. Although for the strict application of the inverse square law and the cosine law the source of light has to be a point, nevertheless in practice, provided the distance of the working surface from the light source is more than five times the maximum projected dimension of the source, the errors introduced will be entirely negligible.

In any practical working interior, the illumination at any point will be the sum of the direct components of illumination from the sources and the light received from the room surfaces by direct reflection and by inter-reflection. This total reflected or 'indirect' component of illumination can be calculated by using energy transfer concepts, or (to a reasonable degree of approximation) by using the BRS 'Split Flux' formula (Hopkinson 1955). The calculation of the direct illumination from a point source requires a simple application of the inverse square and cosine laws and is therefore a matter of elementary mathematics. The illumination from a number of point sources placed at different positions relative to a working surface can each be computed independently and the sum of the separate values of illumination added together to give the total direct illumination on the surface from all the sources. Various computing devices exist to simplify such calculations, and reference should be made to any standard textbook of illuminating engineering.

In practical design problems of illumination calculation, the most frequently encountered difficulty in respect of source size is that of determination of illumination at a point provided by a long row or rows of sources such as may occur with continuous rows of fluorescent lamp fittings. Bellchambers and his collaborators (1968) have developed a method of calculation of both illumination and flux on any plane relative to the axis of a long linear light source using the concept of 'Aspect Factor' and a classification of shape of axial intensity distributions similar to the 'BZ' system discussed later in this chapter.

These devices are useful when detailed information is needed of the *distribution* of illumination over a surface. The lighting engineer, however, is usually concerned with providing a reasonably *uniform* illumination on a large working area from a symmetrical array of light sources mounted at a constant height above the working plane, possibly in the ceiling or suspended at a constant distance from the ceiling. For such a symmetrical arrangement, very simple methods of calculation are available. By far the most frequently used of these methods is the 'Lumen Method'.

The essence of the Lumen Method is to express the average illumination on the working plane, in lumens per unit area, as a function of the total lumens emitted by all the lamps of the installation and of a factor called the 'coefficient of utilisation'. This coefficient of utilisation (sometimes called the 'utilance') is the ratio of that part of the light output of the lamps of the installation which actually reaches the working plane to the total light output. Thus in a completely black room in which radiation falling on all the surfaces is absorbed, the coefficient of utilisation would be determined only by the light which falls directly on the working plane from the lamps. In a practical interior this light is considerably increased by light which falls on the ceiling and walls and is thence reflected to the working plane, or which falls on the floor, is reflected back to the ceiling and walls and thence to the working plane. Thus in practice the total light falling on the working plane is a combination of the direct light from the lamps together with all the internally reflected light which eventually reaches the working plane.

The coefficient of utilisation is therefore:

1. A function of the reflectances of the room surfaces, the higher the reflectance the greater the amount of useful light flux on the working plane.
2. A function of the light distribution from the units so that if most of

the light goes upwards, the coefficient of utilisation will be high only if the ceiling has a high reflectance, whereas if most of the light goes downwards, the ceiling reflectance will have relatively little influence on the flux reaching the working plane.

3. A function of the room dimensions.

The determination of the coefficient of utilisation can be undertaken experimentally by photometric measurements in idealised mock-up rooms in which a wide range of reflectance distributions, light flux output distributions from sources, and room dimensions are examined, from which comprehensive tables can be drawn up; or alternatively the coefficient of utilisation can be computed using a method such as that developed by Jones and Neidhart (1956). Whichever method is employed (and both have been used in illuminating engineering technology) the determination is a sophisticated procedure. For practical use, comprehensive tables of coefficient of utilisation have been published for a very wide range of conditions of room dimensions, ceiling, wall and floor reflectance, and light fitting type and distribution. These tables all assume that the light fittings will be used in the way for which they were designed, and particularly that the ratio of spacing to mounting height above the working plane will be the optimum for the type of light distribution given by the lighting unit.

Until 1961, in order to use the tables of coefficient of utilisation commonly available, it was necessary to assume that the lighting fittings used belonged to one of the broad classifications for which the empirical data had been obtained. However, with the publication in that year of the Illuminating Engineering Society's Technical Report No. 2 (1961) on the British Zonal Method of computing these coefficients, it became possible to classify lighting fittings both by the form of their downward light distributions into ten categories (known as the 'BZ' classification), and by the ratio of the upward to downward light output. For any light distribution which did not fit sufficiently closely into any of the ten mathematical distribution curves, a method of calculating the form of flux distribution, and hence of classification, directly from the polar curve of intensity (the 'Zonal Method') was given.

The application of the Lumen Method enables the total lumens necessary in an installation to give a designed value of working illumination to be computed. The light output of the lamps required must take into account that during the period of the life of the installation, the light output of the lamps themselves will deteriorate and the lighting units will accumulate dirt between routine maintenance operations. If the lamp manufacturer quotes a figure for 'lighting design lumens' in his lamp specification, no further allowance need be made in the calculations for lamp depreciation, but it is usual to make an allowance for other forms of depreciation depending upon the maintenance schedule. The basic formula of the Lumen Method, taking these factors into account, becomes

$$N \times F = \frac{Ed \times A}{CU \times MF}$$

where N = number of lamps required;
F = 'lighting design lumens' per lamp;
Ed = design illumination value;
A = area of working plane;
CU = coefficient of utilisation;
MF = maintenance factor.

It is, of course, impossible to specify a maintenance factor with any great accuracy, and it has generally been customary in approximate calculations to use the figure of 0·8. The IES Technical Report on Lighting Depreciation and Maintenance No. 9 (1967), however, will assist (where it is necessary) the calculation of more accurate values depending on type of fitting, interior, location and cleaning schedule. This report also shows how to calculate the amount by which the illumination will vary during the periods between cleaning and relamping, as well as the most economic type of maintenance schedule.

The calculation of working illumination from laylights, luminous ceilings, and other very large sources of light can be handled by the Lumen Method, as coefficients of utilisation have been worked out for this method of lighting. An alternative procedure is to use methods of flux transfer which have been developed for daylighting technology. These methods are sometimes favoured by architects with skills in daylighting design, particularly if it is necessary to know the distribution of illumination rather than an average value.

Calculation of Luminance

The Lumen Method is primarily intended to permit the calculation of the average level of illumination on the working plane, that is, a horizontal plane at an agreed distance from the floor. For some purposes it is necessary to determine the luminance (brightness) of significant surfaces in the room, particularly the walls and the ceiling. This can be done quite simply by substituting in the Lumen Method formula a 'wall luminance coefficient' or a 'ceiling cavity luminance coefficient' for the value of coefficient of utilisation (Jones and Sampson, 1966). Computed values of these two coefficients are tabulated for a wide range of room dimensions, room surface reflectances, and lighting fitting distributions. Thus:

Average wall luminance = lamp lumens × wall luminance coefficient/wall area.

Directional Characteristics of Lighting

The Lumen Method of lighting calculation evaluates only average quantities, whereas the direct application of the inverse square and cosine laws can be used to obtain the directional characteristics of the lighting; these are sometimes important. Light in a room has directional properties, minimal in rooms with overall distribution of light sources in the roof or ceiling, but highly significant in rooms with a single side window or a single light source. An attempt has been made by Waldram (1954) and by Lynes and his colleagues (1966, 1968) to express these directional effects quantitatively, but at the moment there is no accepted method which can be summarised here at this stage. The problem is discussed briefly later (Chapter 11).

Light Control and Methods of Lighting

The control and directing of light issuing from primary sources is an important branch of illuminating engineering, particularly in situations (floodlighting, roadway lighting, automobile lighting) where the maximum use has to be made of the available light from the source. When artificial light was expensive, interior lighting fittings also had to be designed with precision in order to direct the maximum illumination from the light sources on to the working plane. This is no longer the situation in interior lighting. The use of wall, ceiling and floor surfaces of high reflectance and the recognition that the greatest satisfaction results when the whole interior as well as the working plane is well lit has caused basic changes in interior light fitting design. Whereas formerly the main purpose of lighting fittings design was to make use of reflection and refraction to direct light from the source to the working plane, it is now recognised to be of equal importance to diffuse the source of light in such a way that the luminance of the fitting, as seen by the observer, will be insufficiently high to cause glare, and to allow light from the fitting to be distributed about the room so that a sufficiently high background luminance is provided to reduce discomfort from contrast between the luminance of the fitting and its immediate surroundings.

High room surface reflectances result in these surfaces acting as major factors in distributing the available light flux and ensuring that it reaches the working plane in sufficient quantity. Diffusing materials, either of glass or of plastic, can be full diffusers, that is, the lamp cannot be seen and the diffuser is of uniform, or near uniform, luminance, or they can be partial diffusers, that is, the lamp is seen partially diffused and of lower luminance through the diffuser, while the rest of the diffuser is of a still lower luminance. Experiments on 'contrast-grading' (see Chapter 4) show that in some circumstances partial diffusion may give rise to less glare discomfort than full diffusion. The same optical effect, so far as glare is concerned, can be obtained by the use of suitably designed prismatic and reeded devices which imit the maximum luminance of a fitting as seen by an observer in a normal viewing position in the room.

Reflection of light is still an important method of light control and direction. Reflectors of special contour can be employed to direct light precisely where it is wanted. A paraboloid of revolution with a filament lamp at its focus will, for example, produce a narrow beam of near-parallel light. In interior lighting design, such optical control is often employed for spotlights and for 'down lights' mounted in the ceiling to project light over a narrow area on a picture on a wall, on a table in a restaurant, or anywhere where preferential light is required over a closely defined area. An ellipsoid form can be used where it is required to collect a large proportion of the light from a source, which is placed at one focus, and project it through a small aperture placed at the other. This small aperture may be the film gate of a picture (still or movie) projector, or the small hole in the ceiling used to conceal the lighting of a room, but in the latter case the beam diverges without further optical control as if the hole itself were the source, and this system produces lighting of a large area rather than a sharply defined spot.

Another important use of specular reflecting surfaces is for light control from fluorescent lamps. Such light fittings, mounted in the ceiling, consist of long parabolic mirrors with a tubular fluorescent lamp at the focus. They can be arranged to direct light on to the working plane while the surface of the mirror fitting itself is of very low luminance, and the lamp is screened from normal angles of view. Such lighting systems have been advocated to eliminate all possible glare discomfort in, for example, drawing offices, where high levels of illumination on the working plane are necessary while glare discomfort must be avoided. There are, however, certain visual drawbacks, which are discussed in Chapter 11.

The developments in techniques of pressing, blowing and extruding accurate prismatic patterns on plastic materials has led to a great increase in the range of totally enclosed fittings available with good optical control. It is possible, by directing the light away from the angles at which a fitting is normally viewed, and into angles at which the work and ceiling are preferentially lighted, to produce a lighting fitting which does not produce visual discomfort and which embodies the desirable light distribution characteristics in a clean and simple shape.

Louvers and baffles in various forms are also employed in lighting fittings, particularly in overall luminous ceilings, in order to limit the angle at which the lamps can be seen. In the design of louver systems, it is usual to mount the lamps in a white cavity of very high reflectance, and to place the louvers at a distance below the lamps sufficient to ensure that, with the given lamp spacing, the louver ceiling, as seen by an observer in a normal position in the room, will appear to be of uniform luminance.

Louvers may be of white material, or they may be of grey or coloured surface to reduce their visible luminance or to create a particular colour effect. Alternatively, louvers can be made of specular material of a contour so designed as to give them a very low luminance as seen from a normal position in the room, but to allow the maximum direction of light to a particular area, usually the horizontal working plane.

Polarisation is a further optical phenomenon which is of use in lighting, and is advocated as a means of reducing unwanted reflections from surfaces on the working plane. The advantages to be gained from the use of multilayer polarising devices in ceiling-mounted lighting fittings have led to an adoption of such methods in environments where reflections are unusually troublesome.

Comparative Costs of Lighting

The comparison of the cost of lighting from different sources cannot be made solely on the basis of the power consumed by the lamps for the same light output, nor yet of the difference in lamp life in relation to its cost. The change from, say, filament lighting to fluorescent lighting involves a change in the cost of fittings and in the cost of cleaning them and replacing the lamps.

A mathematical expression to take account of these factors (but ignoring any difference in cost of wiring or fixing the fittings) has been developed, and the derivation of the expression is given here with values which have been made available for various factors. The cost factors will obviously vary with the type of equipment and the economic circumstances of each case, and the appropriate value must be inserted for each particular installation studied.

It is interesting to note that for a planned bulk maintenance and lamp replacing service at the London offices of Esso using fluorescent lighting for Permanent Supplementary Artificial Lighting, the cost of cleaning a three-lamp fitting was found to be 6s 2d per fitting and of relamping 3s 10d per lamp. (Howard, 1967.) These figures may be compared with the cost figures used below of 4s 6d (for a twin 40 W fitting) and 10s 6d, respectively.

1. The basis of comparison is taken as a given light output from the lamps. The efficiency of the fittings or their suitability for providing the amount and kind of lighting required must be considered separately.

2. Assume the necessary quantity of light has been determined to be F total lumens (say 10 000 lumens).

3. Luminous output of lamps is $O_{\text{fluorescent}}$ or O_{tungsten} ('lighting design lumens' or 'average through life') where O_f and O_t are given in Table 2.1. (Say we take 2600 and 1960 lumens, respectively, as an example for 4 ft 40 W White and 150 W tungsten.)

4. Number of lamps required is $N_f = \dfrac{F}{O_f}$ or $N_t = \dfrac{F}{O_t}$ (=3·85 or 5·1).

5. Number of fittings required = n_f or n_t, = $\dfrac{N_f \text{ or } N_t}{\text{number of lamps per fitting}}$

(say 1·98 for double 40 W and 5·1 for single 150 W).

6. Cost of each fitting, including choke and auxiliary but excluding lamps = C_f or C_t. (Say £7 17s 6d fluorescent or 15s tungsten.)

7. Total cost of equipment = $n_f C_f$ or $n_t C$

Table 2.1

LIGHT OUTPUT OF LAMPS

Fluorescent (Lighting design lumens, BS 1853:1967)

Colour	Light output—Lumens *80 W*	*65 W*	*40 W (4 ft)*
North Light / Colour Matching	3100	2700	1700
Daylight	4650	4200	2500
Natural	3600	3100	1950
White (3500°K)	4850	4400	2600

Filament 240 Volts (Average through life, BS 161:1956)

Power (watts)	*Light output (lumens)*
25	200
40	325
40 (coiled-coil)	390
60	575
60 (coiled-coil)	665
100	1160
100 (coiled-coil)	1260
150	1960
200	2720
300	4300
500	7700
750	12 400
1000	17 300

8. Average cost of equipment per annum spread over 15 years

$$= \frac{n_f C_f}{15} \quad \text{or} \quad \frac{n_t C_t}{15}$$

$$\left(= \frac{1{\cdot}98 \times £7\ 17s\ 6d}{15} \text{ or } \frac{5{\cdot}1 \times 15s}{15} = 20{\cdot}8 \text{ or } 5{\cdot}1 \text{ shillings}\right)$$

9. Annual cost of cleaning installation: Found in one particular study to be 4s 6d per fitting for fluorescent and 1s 6d per fitting for tungsten.

$$\text{Cost} = 4s\ 6d \times n_f \quad \text{or} \quad 1s\ 6d \times n_t$$

10. Total annual cost of unlighted fittings

$$\frac{n_f C_f}{15} + (n_f \times 4s\ 6d) \text{ or } \frac{n_t C_t}{15} + (n_t \times 1s\ 6d)$$

11. Cost of lamps l_f or l_t. (By bulk purchase, this might be 8s 7d and 1s $0\frac{1}{2}$d for 40 W and 150 W, respectively.)

12. Cost of replacing lamps r_f or r_t (to include lamp and labour). (Given by one study as 10s 6d and 2s, respectively.)

13. Estimated average burning life of lamp L_f and L_t. (L_f might be 5000 hours.* L_t, according to BS 161:1956, is 1000 hours.)

14. Hours of use of installation per annum = H.

15. Number of replacements of each lamp per annum = $\frac{H}{L_f}$ or $\frac{H}{L_t}$

16. Total cost of lamp replacements per annum

$$=N_f \times r_f \times \frac{H}{L_f} \text{ or } N_t \times r_t \times \frac{H}{L_t}$$

17. Power of lamps = P_f or P_t watts.

18. Power used = $(P_f + p)$, or P_t, where p = power consumed by choke. A poor ballast may consume $0{\cdot}4P_f$, but manufacturers usually claim 0·2 to $0{\cdot}25P_f$. Taking the higher of the manufacturers' figures:

Power used = $1{\cdot}25P_f$ or P_t (in the example, 50 W and 150 W).

19. Cost of power per unit: d pence/kWh.

20. Cost of power for running installation H hours per annum

$$=N_f 1{\cdot}25 P_f \frac{Hd}{1000} \text{ or } N_t P_t \frac{Hd}{1000}$$

$$\left(3{\cdot}85 \times 50 \times \frac{Hd}{1000} \text{ or } 5{\cdot}1 \times 150 \times \frac{Hd}{1000}\right)$$

21. Total cost of running (power plus lamp replacement)

$$=N_f r_f \frac{H}{L_f} + N_f 1{\cdot}25 P_f \frac{Hd}{1000} \text{ or } N_t r_t \frac{H}{L_t} + N_t P_t \frac{Hd}{1000}$$

$$=HN_f\left(\frac{r_f}{L_f} + 1{\cdot}25P_f\frac{d}{1000}\right) \text{ or } HN_t\left(\frac{r_t}{L_t} + P_t\frac{d}{1000}\right)$$

$$=3{\cdot}85H\left(\frac{126}{5000} + \frac{50d}{1000}\right) \text{ or } 5{\cdot}1H\left(\frac{24}{1000} + \frac{150d}{1000}\right) \text{ pence.}$$

22. Total cost of lighting per annum, K = (10) + (21)

$$=n_f\left[\frac{C_f}{15} + 4{\cdot}5\right] + \frac{HN_f}{12}\left[\frac{r_f}{L_f} + 1{\cdot}25P_f\frac{d}{1000}\right] \text{ shillings}$$

$$\text{or } n_t\left[\frac{C_t}{15} + 1{\cdot}5\right] + \frac{HN_t}{12}\left[\frac{r_t}{L_t} + P_t\frac{d}{1000}\right] \text{ shillings}$$

* The length of time for which a fluorescent lamp is run depends on the economic balance or replacement cost *v.* cost of light lost. 5000 hours might well be found economic under some circumstances.

In the example, $N_f = 2n_f$ and $N_t = n_t$.

$$K = n_f\left[\frac{157\cdot5}{15} + 4\cdot5\right] + \frac{H \times 2n_f}{12}\left[\frac{126}{5000} + \frac{50d}{1000}\right]$$

$$\text{or } n_t\left[\frac{15}{15} + 1\cdot5\right] + \frac{Hn_t}{12}\left[\frac{24}{1000} + \frac{150d}{1000}\right]$$

23. Summarising:

Assuming 15 years' amortisation of capital,
4s 6d cleaning cost/year for fluorescent,
1s 6d cleaning cost/year for tungsten,
25% power loss in fluorescent auxiliaries,

we have Ratio of $\dfrac{\text{Annual cost in shillings of Fluorescent}}{\text{Annual cost in shillings of Tungsten}}$ is:

$$R = \frac{n_f\left[\dfrac{C_f}{15} + 4\cdot5\right] + \dfrac{HN_f}{12}\left[\dfrac{r_f}{L_f} + 1\cdot25P_f\dfrac{d}{1000}\right]}{n_t\left[\dfrac{C_t}{15} + 1\cdot5\right] + \dfrac{HN_t}{12}\left[\dfrac{r_t}{L_t} + P_t\dfrac{d}{1000}\right]}$$

where n = number of fittings,
C = capital cost of fitting (shillings),
H = hours of use of installation,
N = number of lamps,
r = cost of replacing lamps (pence),
L = life of lamp,
P = power of lamp,
d = cost of power in pence per kWh.

In the example given, Ratio of Cost Fluorescent/Tungsten,

$$R = \frac{n_f}{n_t}\left[\frac{10\cdot5 + 4\cdot5 + \dfrac{H}{6}(0\cdot0252 + 0\cdot05d)}{1 \quad + 1\cdot5 + \dfrac{H}{12}(0\cdot024 + 0\cdot150d)}\right]^*$$

$$= \frac{n_f}{n_t}\left[\frac{15 + \dfrac{H}{6}(0\cdot0252 + 0\cdot05d)}{2\cdot5 + \dfrac{H}{12}(0\cdot024 + 0\cdot150d)}\right]$$

(to be less than 1·0 for fluorescent to be cheaper).

Values of this ratio are plotted for 4 values of d in Fig. 2.2.

* It will be noted that $\frac{n_f}{n_t}$ is the ratio of the number of fittings ($\frac{1.98}{5.1} = 0.389$ in the example) and the expression inside the square brackets gives the ratio of cost per fitting.

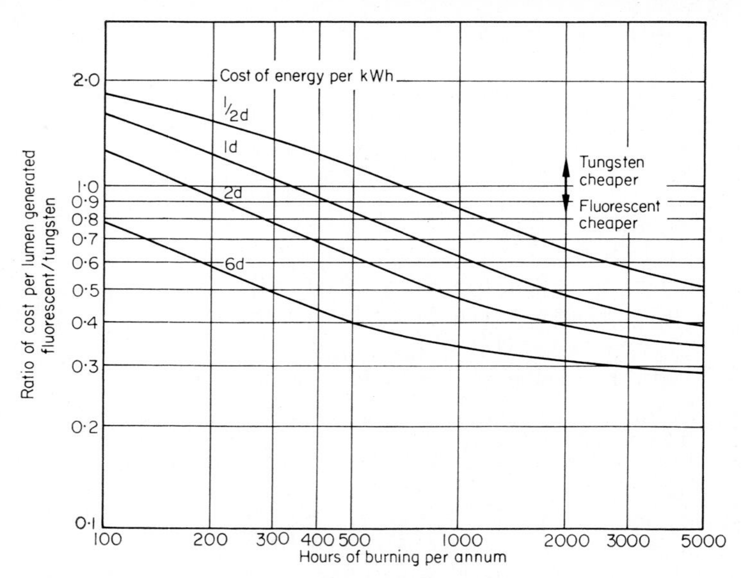

Fig. 2.2 *Ratio of annual cost of fluorescent and tungsten lighting in relation to amount of use and cost of energy*

24. If power costs are based on S shillings per annum per kVA installed plus d' pence per unit:

$$\text{Cost of power} = \frac{N_f 1{\cdot}25 P_f}{1000}\left[\frac{S}{PF} + \frac{Hd'}{12}\right] \text{ or } \frac{N_t P_t}{1000}\left[S + \frac{Hd'}{12}\right] \text{ shillings}$$

where PF = power factor.

$$\text{So that total cost of running} = N_f\left[\frac{Hr_f}{L_f} + \frac{1{\cdot}25P_f}{1000}\left(\frac{S}{PF} + Hd'\right)\right] \text{ shillings,}$$

$$\text{or } N_t\left[\frac{Hr_t}{L_t} + \frac{P_t}{1000}(S + Hd')\right] \text{ shillings.}$$

It will be seen from the graph that only if the flat rate for energy is less than 1d per kWh is tungsten filament lighting cheaper in the installation conditions chosen when the lighting is used for more than about 350 hours per annum. At 2d per kWh, fluorescent lighting is always cheaper if the lighting is used for anything more than ¾ hour per day (180 hours per annum). Individual circumstances may of course vary these relationships.

References

Bellchambers, H. E. *et al.* (1968): 'The Calculation of Direct Illumination from Linear Sources'. Illuminating Engineering Society, London, Technical Report No. 11.

British Standard 161: 1968: 'Tungsten Filament General Service Electric Lamps'.

British Standard 1853: 1967: 'Tubular Fluorescent Lamps for General Lighting Service'.

Hopkinson, R. G. (1955): 'The Indirect Component of Illumination in Artificially Lit Interiors'. *Light and Lighting,* **48** 315.

Howard, J. (1967): 'Lighting Maintenance at the Esso Building'. *Light and Lighting*, **60**, 184–186.
Illuminating Engineering Society, London (1961): Technical Report No. 2: 'Coefficients of Utilisation. The British Zonal Method'.
Illuminating Engineering Society, London (1967): Technical Report No. 9: 'Depreciation and Maintenance of Interior Lighting'.
Jones, J. R. and J. J. Neidhart (1953): 'The Zonal Method of Computing Coefficients of Utilisation and Illumination on Room Surfaces'. *Illum. Engng.* (*New York*), **48**, 141–168.
Jones, J. R. and F. K. Sampson (1966): 'Lighting Design and Luminance Coefficients'. *Illum. Engng.* (*New York*), **61**, 221–229.
Lynes, J. A., W. Burt, G. K. Jackson and C. Cuttle (1966): 'The Flow of Light into Buildings'. *Trans. Illum. Engng. Soc.* (*London*), **31**, 65–91.
Lynes, J. A. (1968) (with C. Cuttle, W. B. Valentine and W. Burt): 'Beyond the Working Plane'. Proc. CIE 16th Session (Washington, 1967), Paper 67.12. Commission Internationale de l'Éclairage, Paris.
Waldram, J. M. (1954): 'Studies in Interior Lighting'. *Trans. Illum. Engng. Soc.* (*London*), **19**, 95–133.

PART II. EXPERIMENTAL STUDIES

3

Visual Performance

The eye as the organ of sight and the brain as the organ of interpretation operate together in the determination of the visual performance of an individual. An individual may have very good sight, that is, his visual acuity may be high so that he can detect very fine detail, but if his interpretative equipment is poor, his visual performance may be low. Alternatively, his sight may be below normal but his intelligence may be high and his skill and experience of such an order that he can overcome his visual incapacity and still register a high degree of visual performance. Young children, for example, often possess a high visual acuity, but their lack of experience in interpreting the world around them results in only a moderate visual performance, while elderly people whose vision has begun to deteriorate can nevertheless perform adequately by virtue of their past experience and their knowledge of the visual world around them.

The measurement of visual performance requires a detailed analysis both of the processes of sight and of the geometric and photometric characteristics of the visual field. We probably know more of the characteristics of the eye as an optical instrument than we do of the characteristics of the brain as an interpretative mechanism. Consequently much of the analysis which goes into the determination of visual performance is concerned with the optics of the situation rather than with its interpretation.

Contrast

An object can only be seen because of its contrast with its background. This is equally true whether it be a printed page, a table-tennis ball (Fig. 3.1) or a black thread on black cloth (Fig. 3.2). Without contrast an object cannot be seen however big (Fig. 3.3)—the mountain vanishes into the mist—or however bright—the silver fox vanishes against the Arctic snow. To be seen, however, an object does not necessarily have to have a different colour or a different reflection factor from its background. If it is a solid object, its form and shape can be revealed by directional lighting (Fig. 3.4), and contained contrast within its outline can also reveal its presence and nature. Even if not a solid object, it may still be possible to arrange the lighting to produce a

Fig. 3.1 *Where there is little or no contrast the outline of the ball is not clearly seen*

contrast between the object and its background (Fig. 3.5(a)) or to reveal its presence by casting a shadow (Fig. 3.5(b)).

Numerical Expression of Contrast

The amount of contrast between an object and its background will govern its visibility, but first it is necessary to decide how to specify contrast quanti-

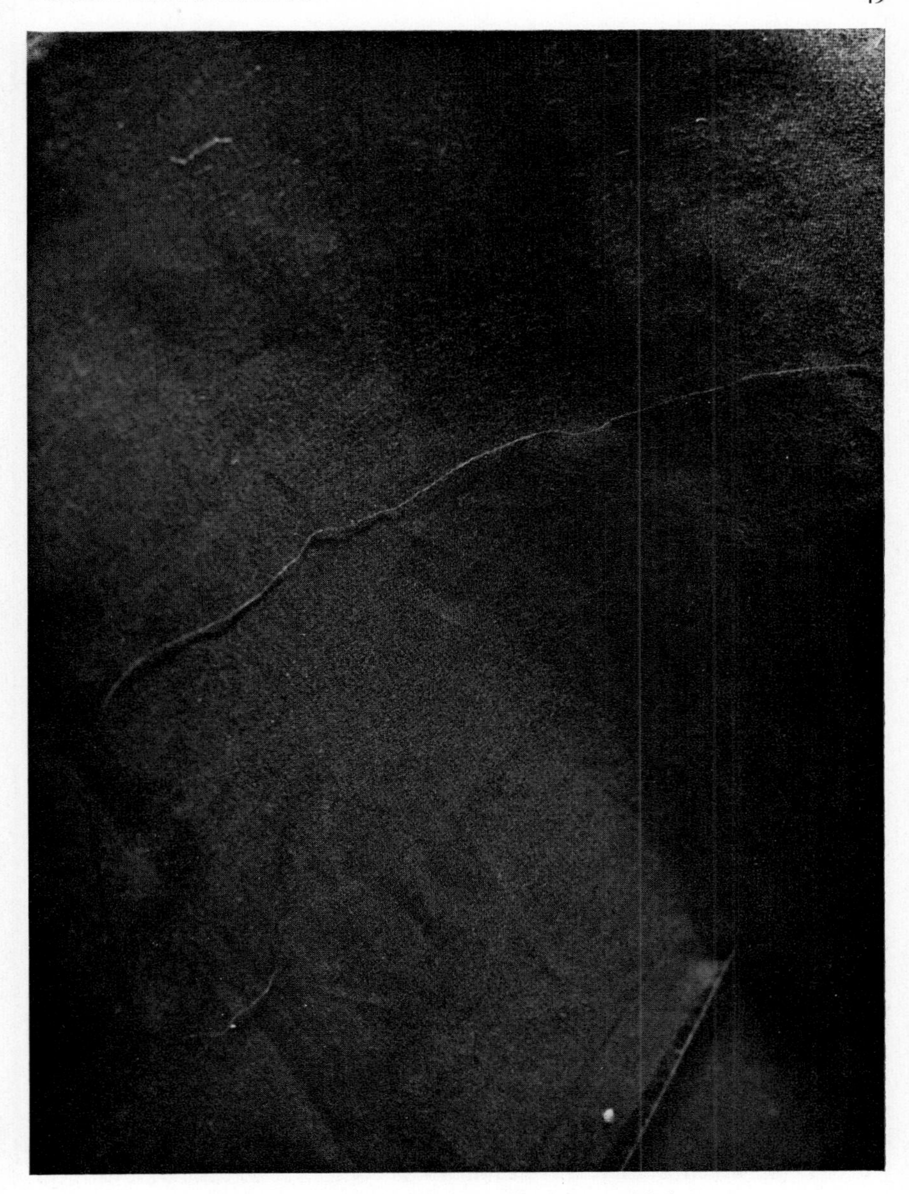

Fig. 3.2 *Contrast enhancement by directional lighting*

tatively. The contrast between an object of luminance B_1 under a given arrangement of illumination, and the background luminance B_0 against which it is seen can be represented in several different ways. Contrast can be specified in terms of *luminance difference*, thus:

$$C = B_1 - B_0$$

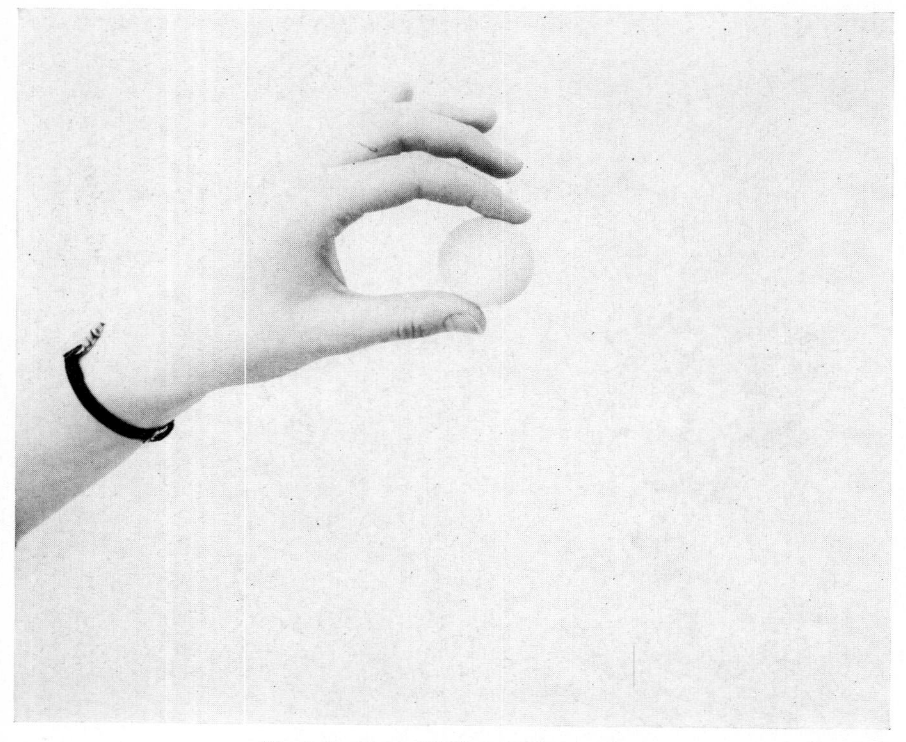

Fig. 3.3 *Poor visibility with low contrast*

Since visual sensations tend to correspond to ratios of stimuli, contrast is usually expressed as a ratio of the luminance difference to one or other of the luminances:

thus $$C = (B_1 - B_0)/B_1$$

or $$C = (B_1 - B_0)/B_0$$

As an example, an object having a luminance of 100 units on a background of 10 units will have a contrast of (100 − 10)/100, that is 0·9. If the object has a luminance of 10 units and the background 1 unit, the contrast is (10 − 1)/10, which is also 0·9. Thus the same contrast, expressed in this ratio form, results from any two luminances of ratio 10 : 1. On this same basis, however, a contrast between a luminance of 100 units and 90 units would be given by (100 − 90)/100 = 0·1, and a contrast between 20 units and 10 units by (20 − 10)/10 = 0·5. Thus the same luminance difference when expressed in the ratio form gives rise to a different value of contrast.

Unfortunately there is no agreement in the literature on the method of expressing contrast. The ratio method is employed most frequently, but in

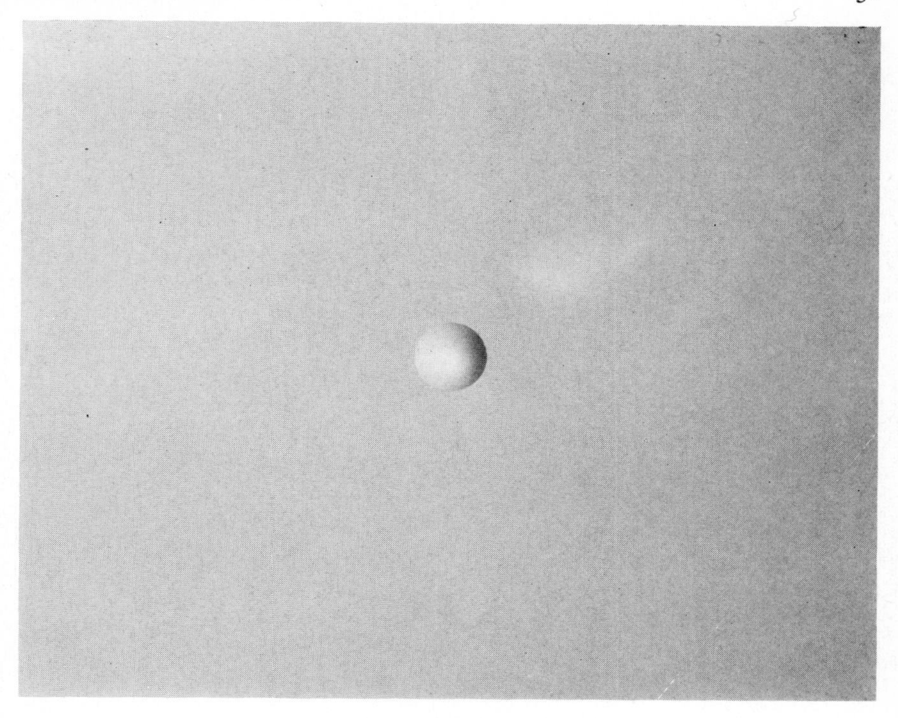

Fig. 3.4 *Form revealed by directional lighting*

different ways. For example, of the authoritative workers in this field, Weston maintained the convention of using the larger value of luminance as the denominator in the expression regardless of whether this larger value was the luminance of the object or the luminance of the background. On this system the value of contrast always lies between 0 and 1. Blackwell and others, however, use the expression:

$$C = (B_t - B)/B$$

where B_t is the luminance of the target (that is, the object) and B that of the background. Hence according to this expression, contrast can vary positively or negatively about zero (i.e. equality of luminance between target and background) up to any value. Thus on this system a target of 100 units on a background of 10 units has a contrast of $(100 - 10)/10 = 9$, whereas a target of 10 units on a background of 100 units has a value of $(10 - 100)/100 = -0{\cdot}9$.

When considering reflecting objects, luminances will be in the ratios of the different reflection factors. Thus a target having a reflection factor of 70% on a white background of reflection factor 80% will have a contrast of $(80 - 70)/80 = 0{\cdot}125$ and if the values are 7% and 8% respectively, the

Fig. 3.5 (a) *Target revealed by luminance difference*

contrast will also be 0·125 (i.e. (8 − 7)/8 = 0·125). Two such contrasts, though equal numerically on the ratio basis, would not, however, have the same subjective effect if seen in the same illumination. The distinction between the darker pair would be less easily seen than between the lighter pair. For a given degree of visibility, the second pair would require ten times the illumination of that of the first. Hence the ratio method of specifying contrast gives

Fig. 3.5 (b) *Low-contrast target revealed by shadow*

no indication of the visibility of the targets unless their luminance is also specified.

Contrast specification in terms of visibility can only be achieved on a basis of apparent brightness (see Chapter 8). Any expression in terms of luminance only is necessarily inadequate. Under some circumstances a better accord with visibility will be obtained by expressing one contrast as a ratio, in others as a difference. The visibility of a target of 100 units on a background of 10 units will be greater than that of a target of 10 units on a background of 1 unit because of the greater level of luminance at which it is seen, and in this case the value of the luminance difference alone may be a better guide to the visual effect. For this reason Guth, Eastman and Rodgers (1953) advocate the use of the luminance differences as the better means of evaluating contrast. In this case the values for the two targets will be 90 and 9 respectively, and although these values do not have any real significance quantitatively as far as degrees of visibility are concerned, they do indicate an order of magnitude in the correct sense.

In general, however, any attempt at a simple expression for the *visual* effect of a contrast between an object and its background in terms of luminance or reflectance or any other *physical* attribute is somewhat meaningless.

Contrast expressed in physical terms is a convention and the method chosen should be based on the arithmetic convenience for the work in hand, and not in any false hopes that the values are a measure of the subjective effect.

Size of Object

The visibility of an object which can be seen by virtue of its contrast will be influenced by its apparent size. The physical size of an object may not be under control, but its apparent size or the angular subtense at the eye may be controlled in that the eye can approach closer to the object and so increase the size of the image on the retina. Alternatively, by the aid of suitable optical devices, the retinal image may be altered in magnitude.

Size and contrast cannot be considered entirely independently. Weston (1961) and Fry and Enoch (1959) have shown independently that contrast acuity and size acuity are interlinked. Weston has shown that an increase of size has a greater effect on improvement of visibility for objects of weak contrast than for those of strong contrast (compare, for instance the bottom

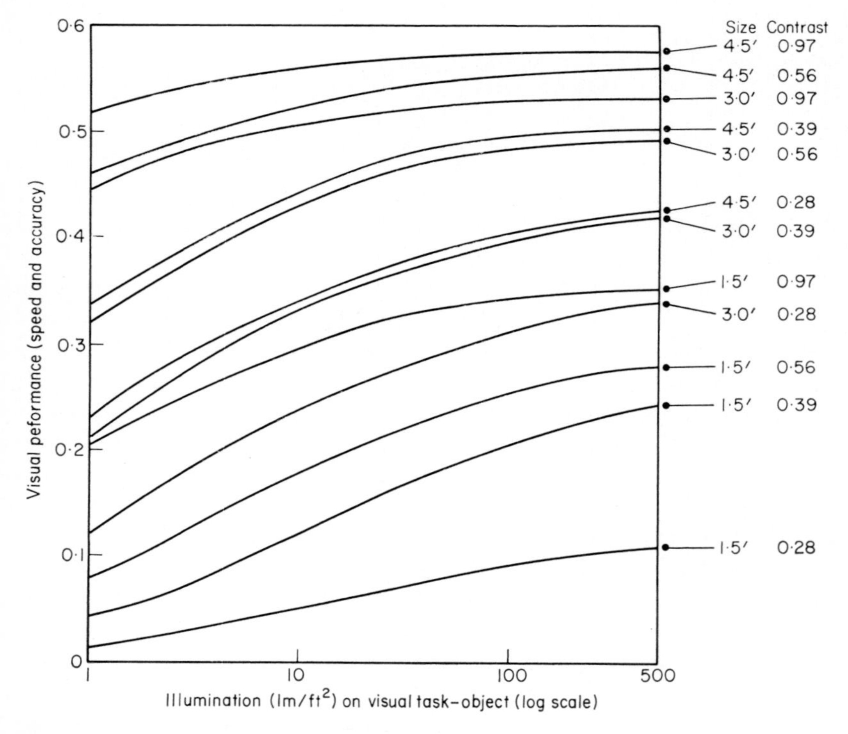

Fig. 3.6 *A family of curves, obtained experimentally, showing the relationship between speed and accuracy of visual discrimination and the illumination of the visual task-object (From Weston 1962)*

curve of Fig. 3.6 with those for $1{\cdot}5'$, $C = 0{\cdot}56$ and for $3{\cdot}0'$, $C = 0{\cdot}28$). No simple relation between size and contrast acuity can be expected, however, since the visibility and recognition of objects is associated with the sharpness of the retinal images in a highly complex manner which is not fully understood.

Controlled Parameters

Factors which may be more readily under the control of the observer are (a) the luminance level at which the object is seen, and (b) the length of time for which it is viewed.

This luminance level may be expressed as that of the object itself, or of its surround, or as a mean value between the two, according to the way in which it is expected that the level of response of the eye is determined. Its value can be expressed numerically, as the product of the illumination on the object and its reflection factor as seen by the observer, and it can be controlled either by varying the reflection factor of the object or its background, or by varying the illumination on the whole task.

Length of viewing time can, in some cases, be altered by the rate at which the printed page or visual task is scanned. This, of course, is not possible if the visual task is presented on a conveyor belt or by some other device at a rate which is determined by considerations other than visibility.

In more complicated tasks, for example where searching is involved, other factors also affect the visibility. Among these factors are the area of the visual field in which the object is likely to be found, and the amount of extraneous information which has to be rejected in the visual field.

Methods of Study

The study of visual performance has been undertaken in two ways, by the observation of the effect of the above factors (size, contrast and speed of vision) on the threshold of perception, or by observation of their effect on the actual performance of a simplified representative task. The work of Blackwell (1946, 1959) is characteristic of the analytical approach while the work of Weston (1945 *et seq.*) (1959) and by Bodmann (1961, 1962) is characteristic of the second method. Work carried out by the present authors (Hopkinson 1946, 1949, 1951) includes studies which may be classified under both these approaches.

A study of the effect of various factors on the threshold visibility is, of course, of great value in providing fundamental data on visual functions, but there are difficulties in applying such data to the visual performance of normal tasks which are usually well above threshold. Assumptions have to be made about the working of the visual mechanism at suprathreshold levels, and use made of more or less empirical factors to translate the fundamental threshold data into information of practical validity. When such a translation is made in order to arrive at a code of recommended illumination values, as has been done in the United States, based on the threshold work by Blackwell, the procedure is always liable to be questioned (see Chapter 9).

Visual Acuity

The standard method of describing the capability of the eye under various conditions is to measure the sharpness of vision, or visual acuity, by which is understood the size of detail capable of being resolved in terms of the angular subtense of this detail at the eye, it being assumed that vision is not limited in any way by low contrast. Visual acuity is usually measured in terms of the size of the detail distinguished in a pattern, or of the critical detail of letters printed in black on a white ground, viewed at a given distance. The pattern usually employed by optometrists for checking visual acuity consists, for example, of letters arranged in rows, each row consisting of letters of a different size. This is the well-known Snellen chart (see Fig. 3.7) which is

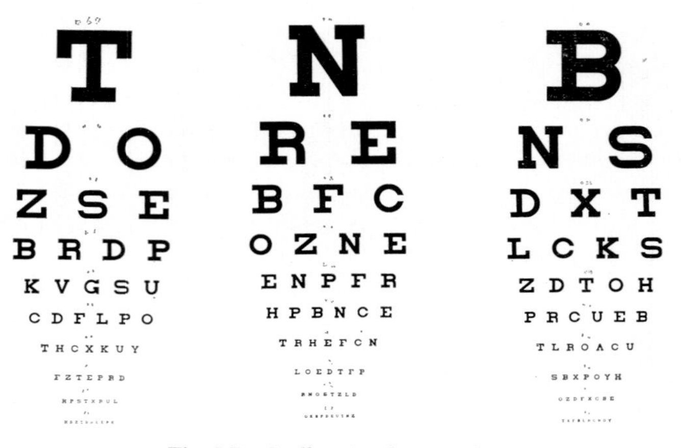

Fig. 3.7 *Snellen visual acuity chart*

usually used at either of the standard distances of 6 metres or of 20 ft. The largest letter on such a chart is usually of such a size that it could be read by a person of normal acuity at a distance ten times that of the standard, i.e. at 60 m or 200 ft. If the person under test, viewing the chart at the standard distance, can see nothing more than this top letter, then his acuity is described as 6/60 (or 20/200) because his vision is so poor that detail which a person with normal vision can see at 60 metres (or 200 ft) can be seen by him only at 6 metres (or 20 ft). In the same way visual acuity of 6/12 means that the line on the chart of such a size that it could be seen by a person with normal vision at 12 metres can be seen by the person under test at 6 metres.

In drawing up the Snellen chart, 'normal' vision is taken as the ability to see one minute of arc as the critical detail of the letter, while the letters themselves subtend an angle, on average, of 5 minutes of arc. This standard is not particularly high and most young people are able to read 6/4·5 and many can read 6/3.

The Effect of Illumination on Visual Acuity

Visual acuity is not independent of lighting conditions. Most people can read more letters and read with greater accuracy from the Snellen chart in a high illumination than in a low illumination. Reading the Snellen chart, in fact, constitutes a simple test for assessing the effect of illumination level on visual acuity. Gilbert and Hopkinson (1949) determined the effect of illumi-

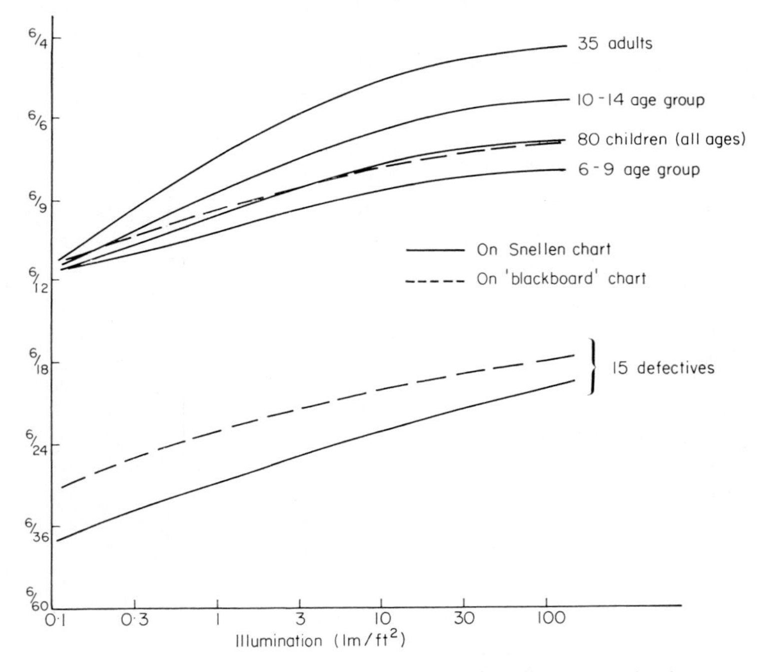

Fig. 3.8 *Mean acuity of samples related to illumination level*

nation on visual acuity using Snellen charts of different types. Among the types selected for this investigation were, in addition to the usual black letters on a white background, white letters on a black background and white letters on backgrounds of different levels of grey. (The investigation was primarily concerned with lighting and vision in schools (Hopkinson 1949) and the white letters on different grey backgrounds simulated effect of white chalk letters on the chalkboard.) In this investigation measurements of acuity were made on 35 adults, and on 80 children of various ages with normal vision, in addition to 15 children whose vision was graded as defective or slightly defective. The relation between visual acuity and the illumination level on the charts for each of the groups and for the normal children divided into two age groups is given in Fig. 3.8. Those with normal vision register an increase in acuity with increasing illumination, but the rate of increase falls off above 10 lm/ft^2 (more markedly, perhaps, for the adults than for the children). On the other hand, this falling off in rate of increase of acuity with illumination level does not appear in the data for the children with subnormal vision, even up to a level of 100 lm/ft^2. This confirms the generally held opinion that people with poor sight benefit more from increased levels of lighting than do people with normal sight, certainly as far as refractive defects are concerned. Some pathological conditions, however, e.g. those involving scattering of light in the optic media, may require that great care is taken when providing increased illumination that the light falling directly on the eye is not increased at the same time.

It is of interest to note the apparent improvement in visual acuity with age. This effect probably represents the improved skill of the older children and of

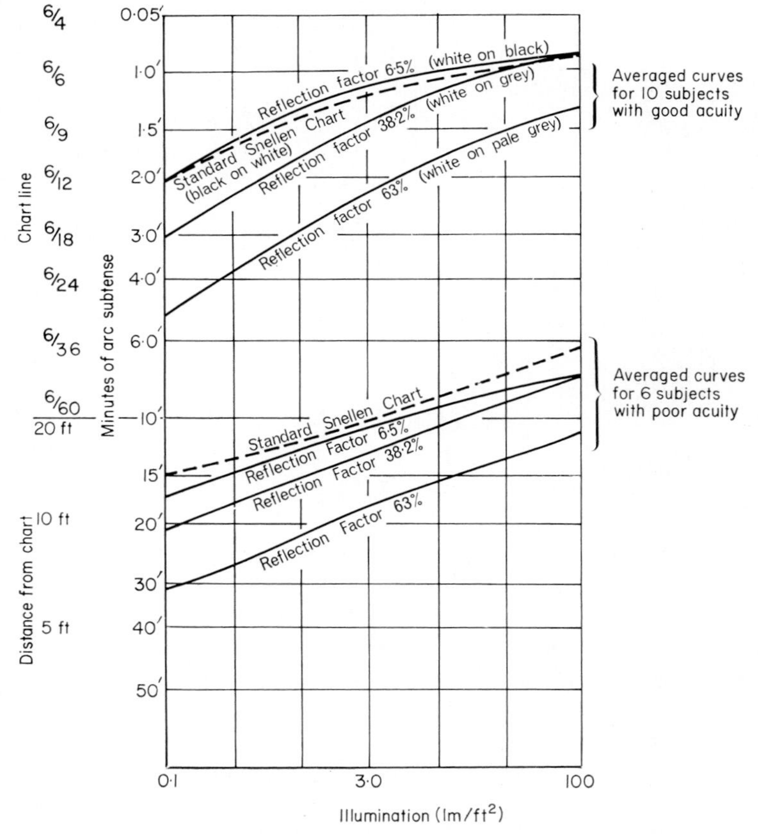

Fig. 3.9 *Average curves showing relationship between illumination and visual acuity assessed on charts with different reflection factors*

adults in guessing the identity of letters indistinctly seen. It can also be seen that there is a distinct advantage in terms of acuity of white letters on black or grey background, for children with subnormal vision.

Effect of Contrast on Acuity

The effect of contrast on acuity was also studied in the course of this investigation. The test charts were variants of the standard Snellen chart, with white letters on a background which varied from a reflectance of 6·7% to 63%. The results of tests using these charts for the measurement of acuity are given in Fig. 3.9. It will be seen that as the contrast is increased by lowering the reflectance of the background from 63% to about 38%, a sharp increase in visual acuity results. This improvement is less, though still present, when the reflectance of the background is further lowered. When the reflectance of the background is 6·5% (very nearly black), acuity is almost double the value with the lightest background.

Effect of Surround Luminance on Acuity

Many studies in visual acuity have been made using a pattern called the Landolt ring (see Fig. 3.10). This is a circle with a gap which can occur at

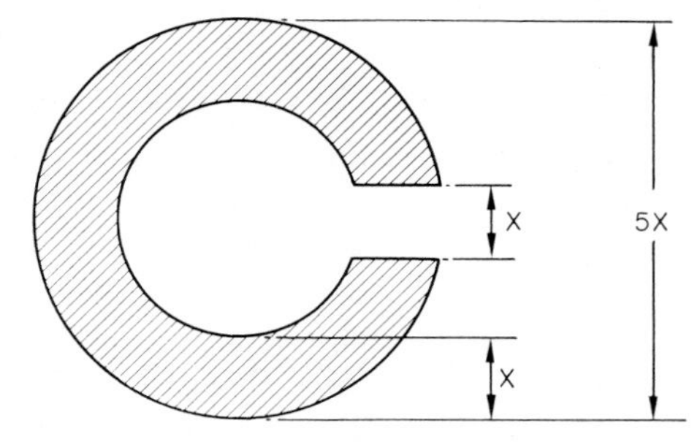

Fig. 3.10 *Dimensions of Landolt broken circle (From Stevens and Foxell 1955)*

any point, and the task usually requires the observer to detect where this gap occurs. A visual task of this type has been used by Lythgoe (1932) in an investigation of the effect of the surround to a visual task on the minimum size of detail in the task which can be resolved. The task was viewed in a surround and the luminance of this surround could be adjusted from a value well below that of the task to a value well above. The results of Lythgoe's experiments for one level of illumination on the task (13 lm/ft^2) are shown in Fig. 3.11, which presents a clear picture of the way in which the recognition of detail in a task is affected by the luminance of the environment in which it is seen. The optimum luminance of the surroundings, or that which gives greatest visual acuity, is rather less than that of the task itself. If the luminance of these surroundings is much less than that of the task, the visual acuity falls

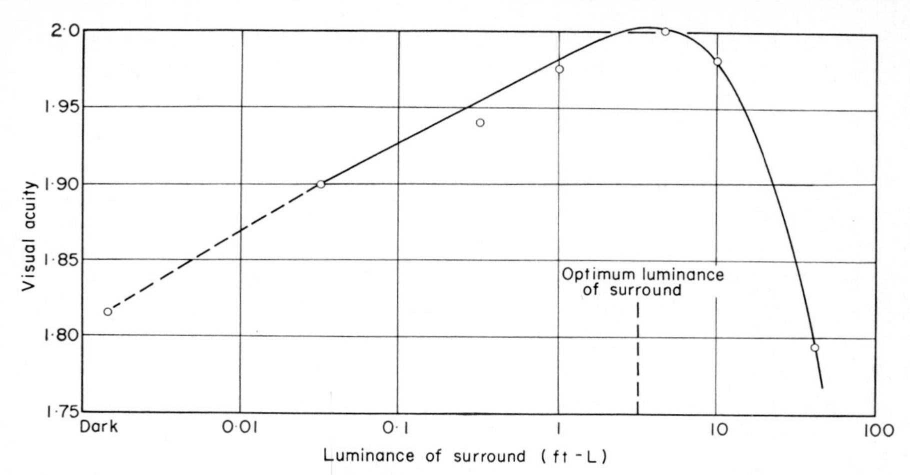

Fig. 3.11 *Modification of visual acuity resulting from changes in the brightness of the surroundings, the illumination on the visual task being constant at 13 lm/ft*^{2} (*From Lythgoe 1932*)

steadily so that with the dark surround it is about 90% of the maximum. If the surround luminance is raised above that of the task, however, the acuity falls more rapidly and is about 90% of the maximum when the surround luminance is ten times that of the task.

The Effect of Surround Size and Luminance on Acuity

The work of Lythgoe has been extended by Stevens and Foxell (1955) who studied the effect of different sizes of surround field at various luminance levels. For equality of luminance of surround and central field, the provision of a surround field (around the 0·5° work point) appeared to have little effect on acuity when the luminances of task and surround were low (Fig. 3.12). At higher levels, however, very marked increases were produced by relatively small sizes of surround field. There is little difference in effect between a surround field subtending 38° and one of 120°. The bulk of the surround effect appeared to be achieved by a surround of 6° in extent. The investigation confirmed the finding of Lythgoe that maximum acuity is obtained with a surround luminance equal to or slightly lower than the task area luminance, but these authors point out that the general rise of the threshold curve as the surround luminance is brought up to approximate equality with the field luminance implies that efforts to raise the luminance of large areas of the surrounding field right up to approach that of the task area are not always justifiable.

Ease and Speed of Reading

The investigation by Gilbert and Hopkinson was extended (Hopkinson, 1949) in a manner which was intended to give a more direct relationship

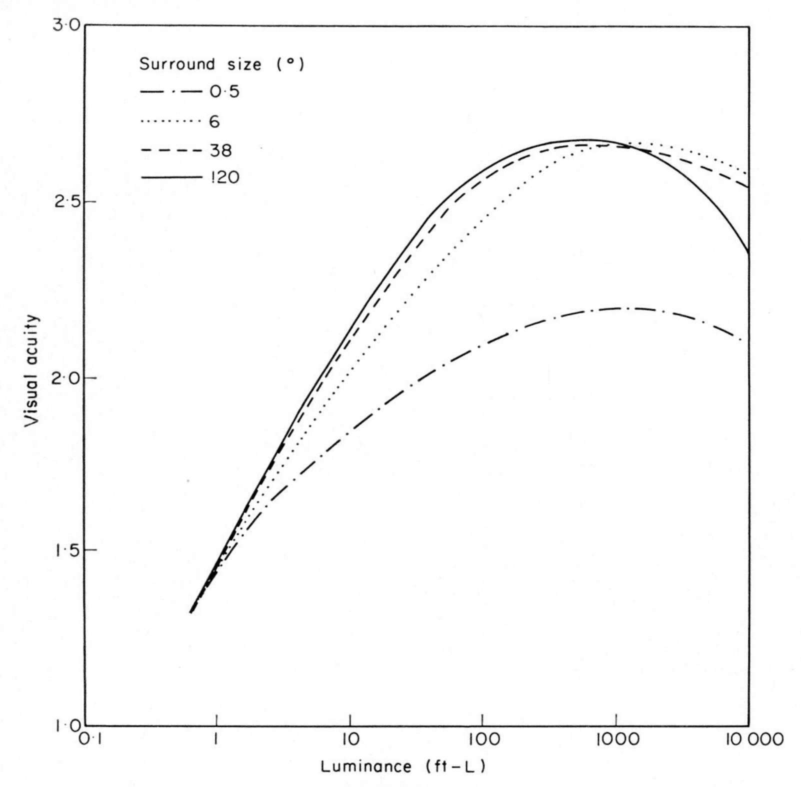

Fig. 3.12 *Relation between visual acuity and luminance for various sizes of surround (luminance of surround same as that of central field) (From Stevens and Foxell 1955)*

to practical levels of performance at which visual tasks are normally carried out. Reading matter set in type was used as the visual task, and the observer was asked to recognise a series of criteria of readability while at the same time his speed of reading and the errors which he made were recorded by the experimenter. The criteria employed were 'just recognisable' (indicating that the observer could just see that the task consisted of print), 'barely readable', 'reasonably readable', and 'easily readable'. The size of the print used for this visual task corresponded approximately to the 6/5 line on the standard Snellen chart. Charts with black letters on a white background of 75% reflectance and with white letters on a grey (15·5% reflectance) and black (7·5% reflectance) background were employed in this study.

Some of the results of this investigation are shown in Fig. 3.13. It is apparent that the relationship between illumination and readability is very much the same as that between illumination and acuity, in that a 'law of diminishing returns' appears to operate above about 30 lm/ft^2. It will also be noted that at low levels of illumination, a change of about ten times in illumination is generally necessary to register a change from one criterion to the next

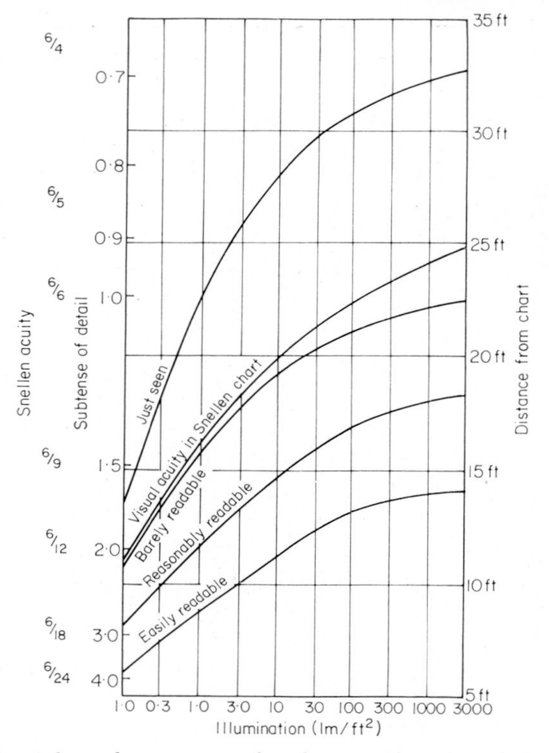

Fig. 3.13 *Relation between ease of reading, visual acuity and illumination*

(for example, from 'barely readable' to 'reasonably readable'). An equivalent change in readability results from increasing the size of the detail by about one half, or alternatively by approaching the task to about three-quarters of the previous distance. These findings are, of course, of great practical significance, since they indicate that in the design of school classrooms, for example, gains in visual performance which require very large increases in the amount of light on the chalkboard can equally well result from quite small changes in the distance of the children from the chalkboard, or in the size of the letters used.

The Effect of Contrast

Readability is improved by greater contrast, but of particular interest are the preferences expressed by observers. At low levels of illumination (0·1 lm/ft^2) 65% of the observers preferred the white on black chart, as compared with 30% preferring the white on grey, and only one observer preferring the black on white. At 10 lm/ft^2 all three charts were equally satisfactory. At 100 lm/ft^2 the black on white chart was now preferred by 50% of the observers while, with illumination values above 100 lm/ft^2, the preference of the white on black chart continued to fall. In full sunlight (3000 lm/ft^2), black on white and white on grey were equally preferred (40% preference each), while only 20% preferred the white on black chart.

Since this work was done, more evidence has accumulated of the advantages of white lettering on a darker background for signs and information charts which may be used in low levels of illumination. It should be noted that the preferences are by no means overwhelming, however, and certain types of information may not be best served in this way. It is often necessary to investigate individual cases on their merits. There is also some evidence, anecdotal rather than experimentally confirmed, that elderly people, and especially those in the early stages of cataract, who suffer from a greater degree of light scatter in the optic media, see better with white letters on a dark ground.

Visual Acuity and Ease of Reading

A conclusion of particular interest from the study by Gilbert and Hopkinson was that people who had the same standard of visual acuity do not necessarily perform visually with the same ease. Several of the subjects who had 'normal' vision gave widely different assessments of ease of reading of the same material. Several check tests were done on these observers, and the results indicated quite clearly that individuals who have visual acuity classified as 'normal' may, in fact, vary widely in the difficulty which they encounter in an ordinary reading task. The reason for this is not clearly understood, although in extreme cases it can be shown that people with this reading difficulty tend to read word by word rather than by large saccades taking in several words at a time. This is a point of great importance to bear in mind when considering the visual health of school children and of industrial workers. The Snellen acuity, as determined in an ordinary standard optometric examination, is clearly insufficient for the purpose of evaluating visual performance under educational and industrial conditions. The condition known as 'dyslexia' (word-blindness) is now an accepted clinical condition, but these studies suggest that mild degrees exist in 'normal' observers.

The Effect of Luminance of Surroundings on Ease of Reading and Comfort

In a further investigation, Hopkinson (1949, 1951) extended the investigation on subjective visual performance to cover the effect of the luminance of the immediate surround of a task on the ease with which it could be performed. The object of this study was to complement the work of Lythgoe by investigating conditions at levels of visibility above the threshold conditions of Lythgoe's experiments. In Hopkinson's study subjects were given a visual task consisting of reading part of a telephone directory surrounded by an area whose luminance could be adjusted in relation to the average luminance of the task itself. The subject was asked to increase the luminance of the surround from equality with that of the task and so to adjust the ratio between these two luminances to achieve, in turn, each of three clearly defined criteria of visibility of the task, 'just easily visible', 'just readable', and 'just barely legible'. Four criteria of discomfort which arose from the excess in surround luminance over the luminance of the task were also determined by the subject.

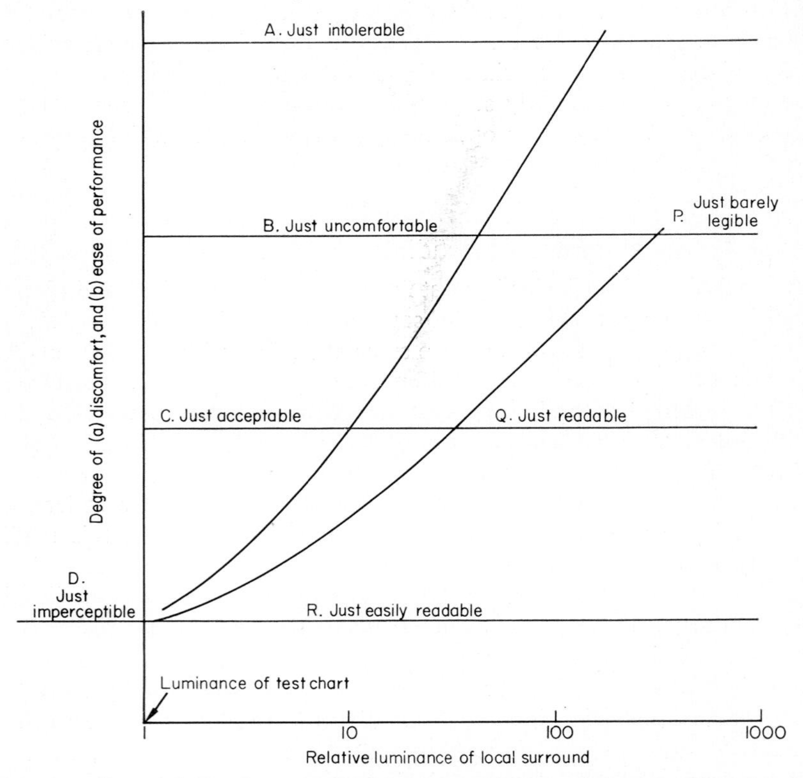

Fig. 3.14 *Effect of the brightness of the local surroundings to the visual task on (a) visual comfort, and (b) ease of visual performance*

The relationship between the criteria of ease of performance and of discomfort is indicated diagrammatically in Fig. 3.14. This suggests that visual comfort is affected more by the increase of luminance of the surround area than is the performance of the task itself. An increase in surround luminance to about thirty times that of the task reduces the level of performance of the task from 'just easily readable' to 'just readable', but the discomfort is changed from 'just imperceptible' to 'just uncomfortable'. Unfavourable conditions of surround luminance are therefore more likely to give rise to complaint of glare discomfort, while the effect on performance or readability may not be great.

Luminance Threshold Studies

Investigations such as those by Hopkinson and his colleagues on supra-threshold conditions suffer from the need to rely upon the subjective assessments of observers. Skilled observers are known to be able to make reliable and repeatable observations, but the variance in the observations is considerable. There is still a body of opinion which prefers to work with thresholds, which can be assessed with less variance, and to translate this threshold information into conditions corresponding to practical situations by means of

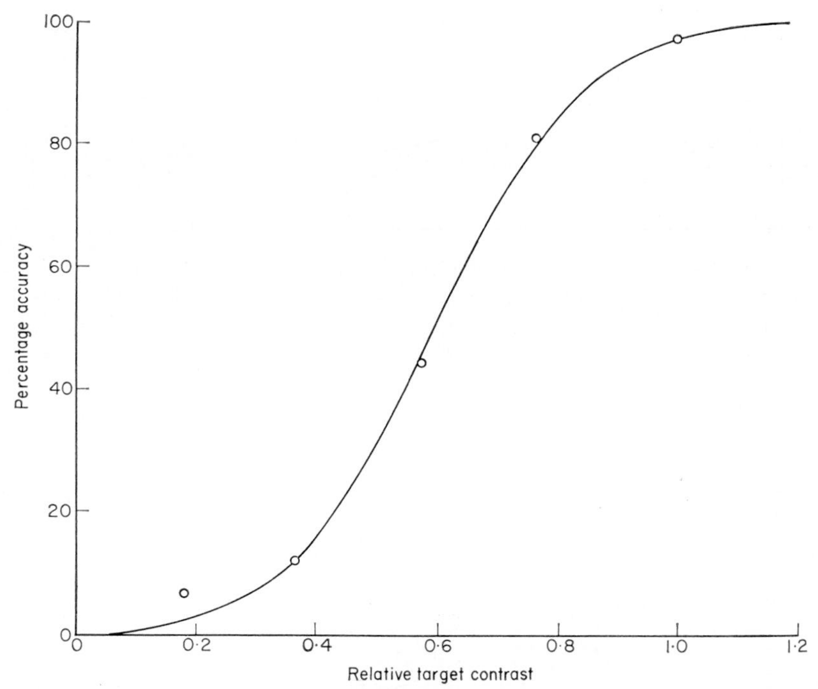

Fig. 3.15 *Sample accuracy curve: response probability data fitted by a normal ogive (From Blackwell 1959)*

some empirical field factor. In the extensive studies on threshold contrasts by Blackwell (1959), the threshold of detection was determined by presenting a luminance difference at the centre of a field of uniform luminance over at least 30° in all directions from the central fixation point. The luminance differences were presented in the form of transilluminated discs of various sizes, varying between 0·8 to 51·2 minutes of arc subtended at the eye. These targets were illuminated for a period of time varying from 0·001 second to 1 second. The subjects (two in all) were told the size of target to expect and the duration of its exposure. The position in which it would appear was closely defined and also the moment during one of four possible periods of

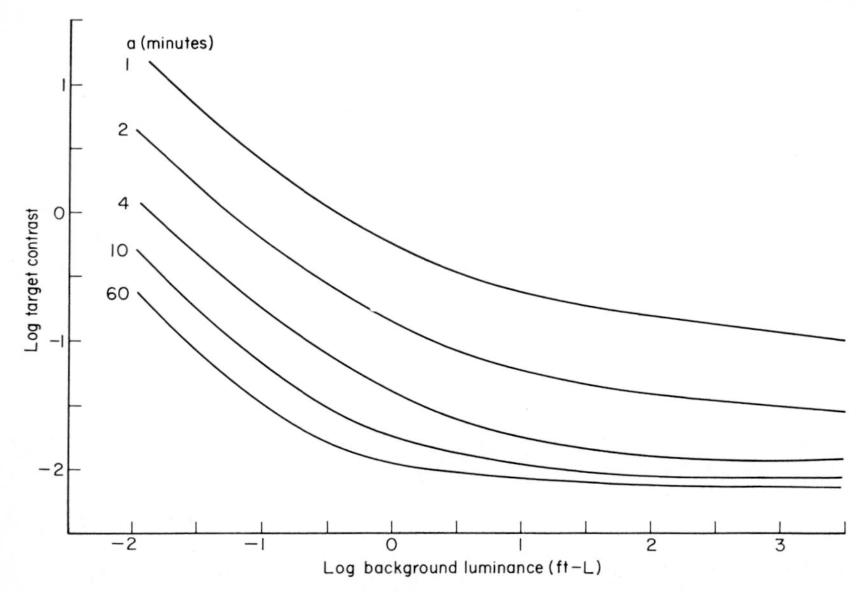

Fig. 3.16 *Smoothed threshold contrast curves for a one-second duration (50% accuracy) (From Blackwell 1959)*

Each curve represents a standard disc target whose diameter *a* is indicated in minutes of arc

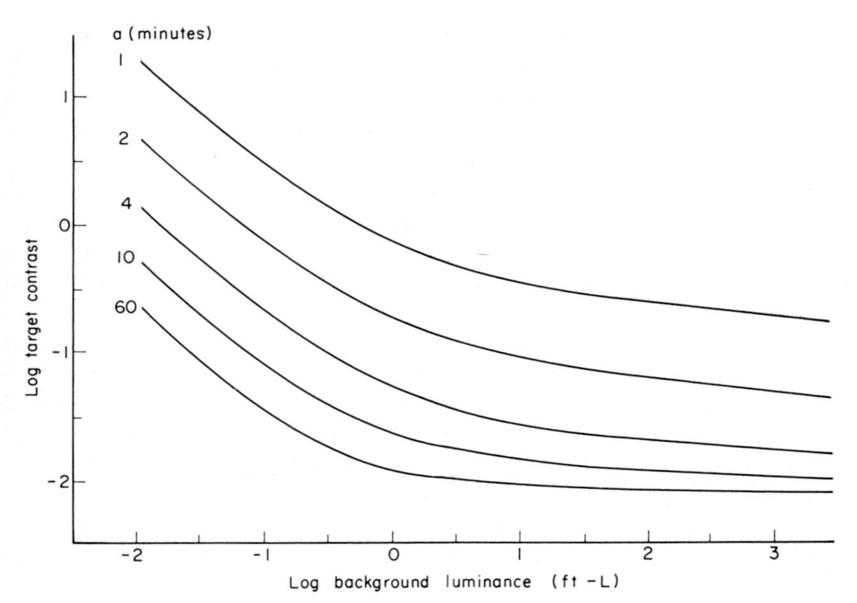

Fig. 3.17 *Smoothed threshold contrast curves for a 1/3-second duration (50% accuracy) (From Blackwell 1959)*

Each curve represents a standard disc target whose diameter *a* is indicated in minutes of arc

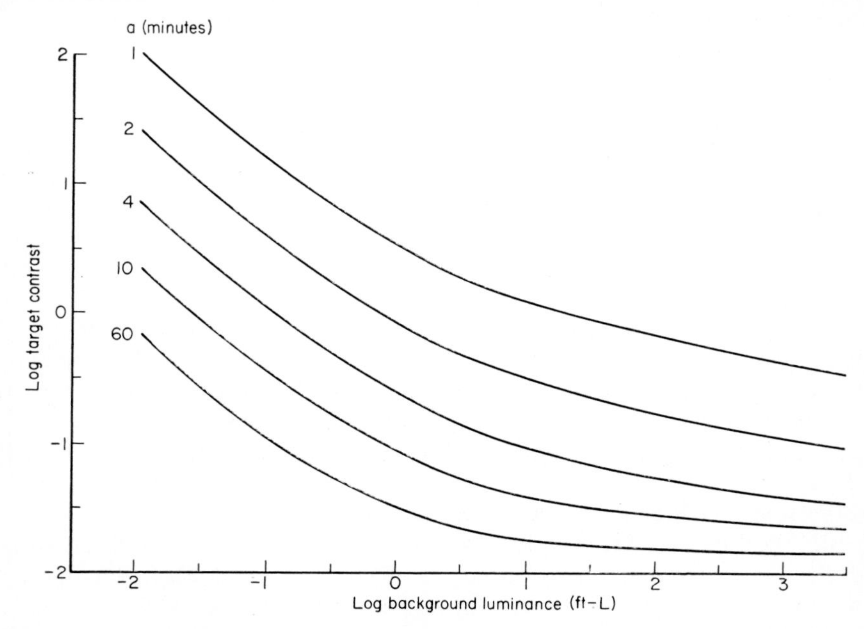

Fig. 3.18 *Smoothed threshold contrast curves for a 1/30-second duration (50% accuracy) (From Blackwell 1959)*

Each curve represents a standard disc target whose diameter *a* is indicated in minutes of arc

time when it would appear. The subject was left to decide in which of the four periods of time it actually did appear. This period and the luminance of the target were controlled by an automatic presentation device.

Since the subject had no control of the characteristics of the target, it was necessary to arrange that the target was presented at a number of values of contrast which were near (slightly above and slightly below) the threshold. The number of times a target was correctly seen or failed to be seen, after making a suitable correction to allow for possibility of guessing, provided data which was subjected to probit analysis in order to fit a normal ogive curve such as is shown in Fig. 3.15. The normally accepted definition of threshold value of target parameter is that at which there is a 50% probability of the target being detected. It is evident, however, that various other levels of probability could equally well be taken, for example, 99% probability of detection (but excluding 100% certainty, as the ogive is asymptotic to this value). Figs. 3.16, 3.17 and 3.18 show the threshold curves in terms of limiting target contrast against background luminance for three presentation times.

It is from these data that Blackwell has deduced his relative values of contrast for different levels of accuracy of seeing, and, by inference, of visual performance.

The speed factor in visual performance is assessed by the length of time for which the threshold stimulus is exposed, and by a further inference, the number of items of information which can be assimilated in one second is assumed to be the inverse of the duration of one threshold stimulus.

Relation of Threshold Data to Visual Performance

The conversion of threshold data of this type into practical performance data, in order to specify illumination requirements for tasks of various difficulty, requires some form of translation. This translation takes the form of a number of steps, some of which may be easier to validate as strictly logical than others.

The most important of these translation steps is the introduction of what Blackwell calls a 'Field Factor' to allow for the fact that in the laboratory the position in the field of view and the instant at which the target will appear are closely defined for the benefit of the observer, and the distance governing the amount of accommodation and convergence required is also fixed. Moreover, the observer has to make a forced choice in which even vague impressions of seeing can count as success.

Blackwell carried out experiments to determine the effect of the first two of the above factors. Furthermore, he devised a 'Field Task Simulator' to show the comparison between a practical inspection task and a laboratory threshold measurement.

Field Task Simulator

In the Field Task Simulator a series of fifty circular plaques of diameter 4 in passed in front of the observer at a speed which could be varied so that the whole fifty plaques could be inspected in only twenty seconds, or in as much as fifty seconds. The plaques could be transilluminated individually from below to give a spot of light of a size and contrast with the rest of the plaque which could be varied, but usually only two or three plaques were so illuminated. The task of the observer was to detect which these were and to signal accordingly with a push button.

The results for two different rates of working are shown in Figs. 3.19 and 3.20. It will be seen that over a range of about 10 : 1 in contrast, the contrast curves can be represented reasonably closely by threshold curves multiplied by constant factors of 6·0 and 7·25. Blackwell has suggested that different types of task may require different values of Field Factor to allow for different characteristics encountered in performing these tasks. For general use, he has advocated a Field Factor value of 15.

It has been argued that this type of field experiment is still essentially one involving threshold contrasts, and it would be surprising if it did not give curves parallel to the threshold curves. Blackwell's experiment does not, therefore, answer the basic question, of the relationship between threshold visibility and practical tasks which are mostly well above threshold. Blackwell's critics therefore say that the Field Task Simulator gives no information

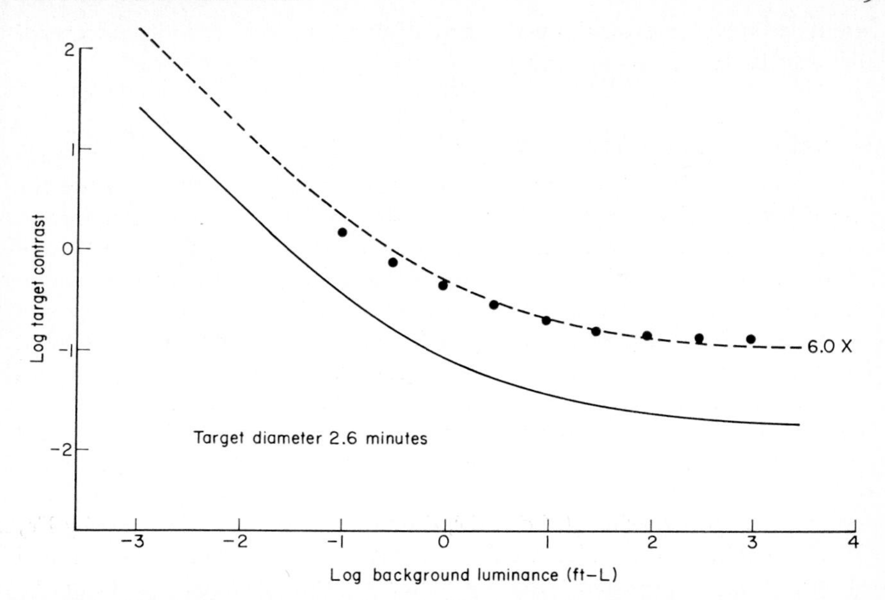

Fig. 3.19 *Threshold contrast data from the Field Task Simulator, for a rotational speed yielding plaques at the rate of one per second (50% accuracy) (From Blackwell 1959)*
The solid curve represents a threshold curve interpolated from the basic laboratory data. The dashed curve represents the laboratory contrast values multiplied by a factor of 6.0

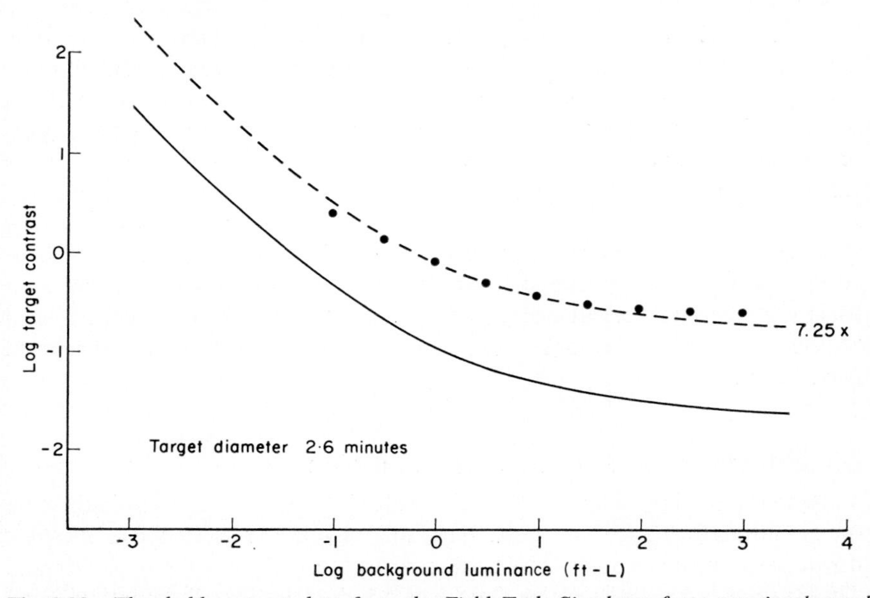

Fig. 3.20 *Threshold contrast data from the Field Task Simulator for a rotational speed yielding plaques at the rate of 2.5 per second (50% accuracy) (From Blackwell 1959)*
The solid curve represents a threshold curve interpolated from the basic laboratory data. The dashed curve represents the laboratory contrast values multiplied by a factor of 7.25

about the speed or ease with which actual suprathreshold tasks such as reading clear print can be performed.

Visual Task Evaluator

Blackwell made one further series of experiments in order to validate the use of his extensive threshold data for evaluating the visual performance of tasks which are easily visible. He designed an apparatus called the 'Visual Task Evaluator', the principle of which is to reduce the visibility of an actual task by reducing its contrast by a known amount. The contrast of a standard target of a given size, which, when subject to the same amount of contrast reduction, is also reduced to threshold visibility, is then determined, and this target is then said to be the equivalent of the actual visual task.

It will be realised that the validity of Blackwell's proposal rests on the assumption that, if two tasks are equivalent at threshold visibility, they are then also equivalent in terms of visual performance at suprathreshold levels of contrast. His experiment, however, does not prove this, but work by Fry (1962) maintains that Blackwell's assumption is justified. Luckiesh and Moss (1934) developed a 'visibility meter' which depended for its operation upon exactly the same principle, that is, of reducing the contrast of a practical task to threshold, the 'meter' being calibrated against a standard visual task also reduced to threshold. Dunbar (1940) and Hopkinson (1946), by two independent methods, showed the dangers of this assumption.

Nevertheless, Blackwell's work has received support, particularly from the Illuminating Engineering Society of America, which has used it as the basis for its recommended levels of illumination in working areas. If Blackwell's assumption about the relation of threshold to suprathreshold visibility is accepted, his threshold probability ogive curve can be used to predict percentage of 'performance' (that is, perception of the details of the task) under various levels of illumination. The development of this into a means of specification of lighting levels is discussed later in Chapter 9. The Commission Internationale de l'Éclairage agreed at its Washington Session (CIE 1968) to adopt Blackwell's proposed standard relative threshold contrast curve as a basis for visual task evaluation, and to study the design of psychophysical 'visibility meters' as a means of expressing suprathreshold tasks in threshold parameters.

Direct Methods of Performance Evaluation

Beuttell (1934) proposed a method of studying the effect of illumination on the actual performance of a task. Beuttell's proposal was to break the task down into what he called 'critical contrast' and 'critical size of detail'. This system of critical task analysis was adopted by the Illuminating Engineering Society of Great Britain as the fundamental basis for its recommendations on lighting levels, following an extensive experimental development of

Beuttell's proposals by H. C. Weston (1945). Beuttell's concept of the critical size of detail was the angular subtense at the eye of the detail which determined whether or not the task could be accomplished; and the critical contrast was the contrast which again determined whether or not the task could be performed. For example, in distinguishing the word 'foot' from the word 'feet', the critical detail is the recognition of the presence of the gap and the horizontal bar in the letter 'e', which are both absent in the letter 'o'. Beuttell elaborated this matter of critical detail and critical contrast in relation to practical visual tasks. Weston developed this proposal experimentally and performed a large number of studies which have subsequently been the basis for much thinking on the problem of visual performance.

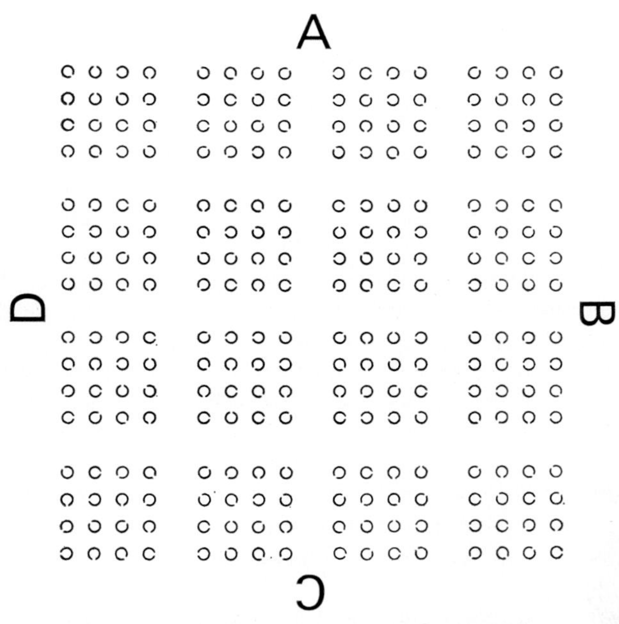

Fig. 3.21 *Weston Landolt Ring performance chart*

Weston developed a series of visual tasks which consisted of a large number of Landolt rings printed in a pattern (see Fig. 3.21) arranged so that the positions of the gaps in the rings fall in a random distribution in the pattern. Charts of this kind were printed with different contrasts, by varying the reflectance of the paper on which they were printed. Landolt rings of different sizes were made available so that both contrast and critical detail (determined by the size of the gap in the Landolt ring) could be determined over a sufficiently wide range for the investigation.

The task of the observer was to cancel, with a pencil, every ring with the gap in a particular specified direction. In order to do this, he had to search through the whole array of rings in a systematic way, the method being determined by himself. Performance therefore depended on both accuracy and speed, exactly as is required by the majority of industrial and office tasks. Weston's method of scoring took into account the time taken in the purely nonvisual part of the task. He did this by setting a separate test which involved the cancellation of a similar number of rings regardless of the orientation of their gaps, and he subtracted this 'mechanical time' from the total time taken by the observer during the visual task. The remainder was taken by Weston to represent the time for the purely visual part of the performance. He then detailed the relationships between the illumination on the Landolt ring tasks, the critical size and the critical contrast, expressed in terms of the visual performance achieved. Weston's extensive experimental data should be consulted, but a summary of his basic relationships is shown in Fig. 3.6.

It will be seen that visual performance increases steadily as the illumination is increased up to a value of about 30 lm/ft^2 (300 lux) for the more easily visible tasks (that is, larger size and higher contrast), but that beyond this value the increase falls off. For tasks of smaller critical detail, the performance increases steadily up to about 300 lm/ft^2 (3000 lux). Beyond this value, improvement in visual performance with increased illumination falls off steadily.

This levelling of the performance curve at high values of illumination indicates that there would appear to be some level of illumination on a task (and therefore luminance of a task) beyond which no further advantage will accrue, and that this level would appear to be of the order of the level of diffused daylight on an overcast day in the open.

Examination of Weston's data (Fig. 3.6) shows also that tasks which differ appreciably in both contrast and size cannot be performed equally well merely by raising the illumination on the most difficult, that is, while high illumination can improve visual performance on difficult tasks as well as on easy tasks, there is a very definite limit to the value to be obtained from increased lighting levels.

Weston (1961) proposed the use of a nomogram which would enable the necessary levels of illumination to be determined for any visual task whose critical detail and contrast could be obtained. This nomogram (Fig. 3.22) was adopted by the Illuminating Engineering Society of Great Britain in the 1961 edition of its code of recommended lighting practice. Weston's experimental data on his Landolt ring charts are here assumed to apply to any visual task no matter how different in appearance from a Landolt ring, taking as the basic assumption Beuttell's proposal that the illumination necessary on any visual task would be fully prescribed by the critical detail and contrast. Weston and his co-workers (1963) have provided some justification for this assumption in their study of more complex task patterns.

The development of the British IES Lighting Code from Weston's data is discussed more fully in Chapter 9. The Beuttell-Weston proposals, the experimental development by Weston, and the translation into recommended

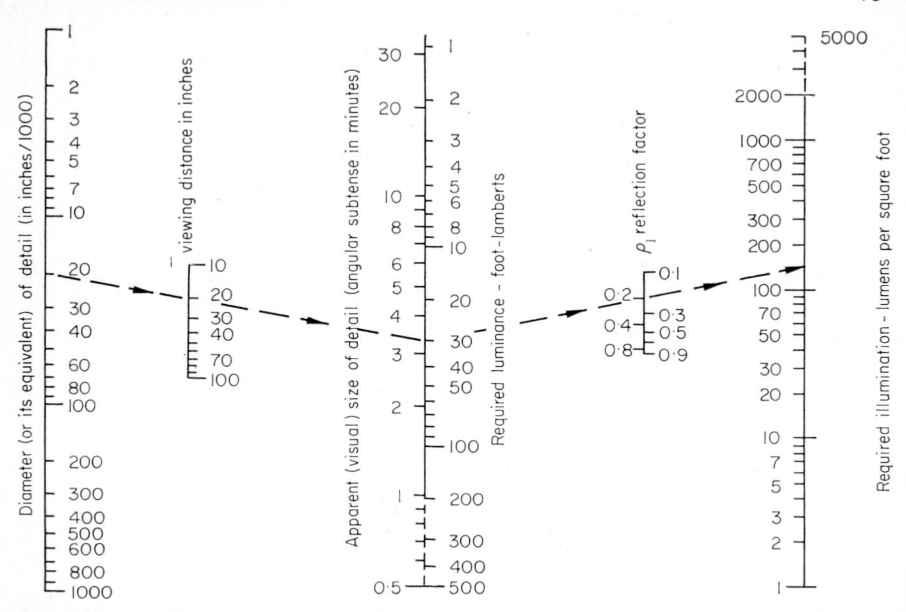

Fig. 3.22 *Nomogram based on the derived relation between size and luminance; the illumination required, according to the size and 'lightness' of the detail that needs to be seen, is found as exemplified in the diagram* (*From Weston 1961*)

levels of lighting practice by the British IES have received considerably less criticism than the comparable work of Blackwell in the United States, but this may be due partly to the fact that the recommendations which have been put forward in Britain are less prodigal of light than those in the United States.

Nevertheless, it is necessary to point out that no field validation of the recommendations for illumination levels on practical visual tasks has ever been attempted either for Weston's work or for that of Blackwell. No experiments have ever been done to demonstrate that, on practical visual tasks, the provision of the level of illumination recommended by either of these two methods does, in fact, result in the visual performance predicted from the basic experiments. Until this field study is done, lighting codes based on either Blackwell or Beuttell-Weston cannot be considered to be fully validated. The matter is of interest from a scientific standpoint, but in practice the recommended levels of illumination obtained by either method are so high and so far beyond the minimum necessary for safe and minimal seeing that the argument is of economic rather than of academic consequence.

Search Tasks

Bodmann (1961, 1962, 1967) devised a method which was a direct development in principle of that of Weston, but which involved the measurement of

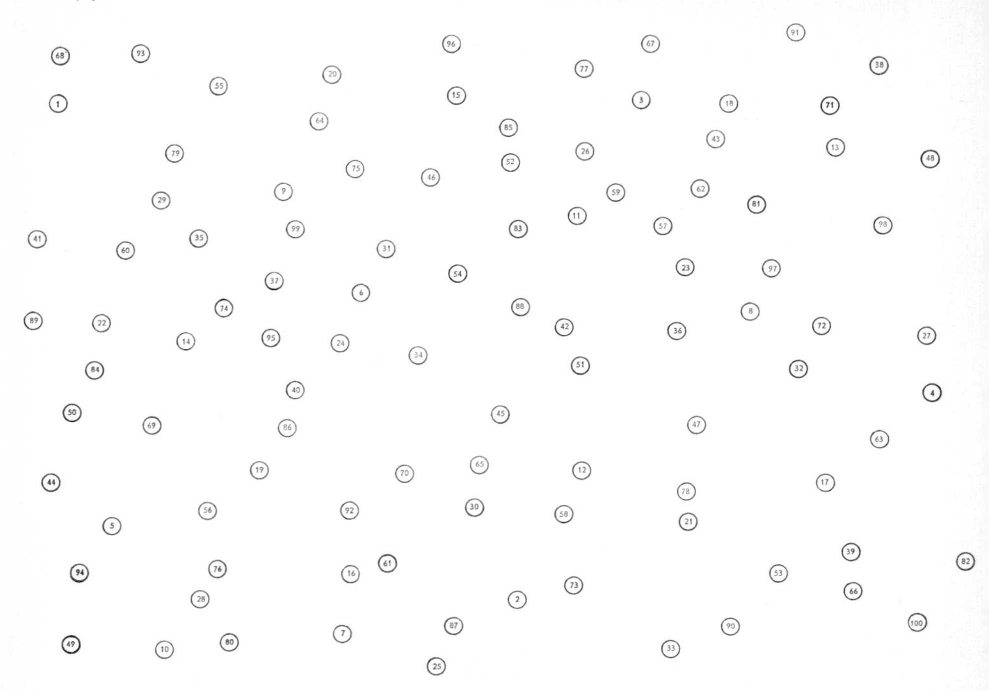

Fig. 3.23 *Bodmann search-task chart*

speed of performance for 100% accuracy, and his experiment also combined objective measurements of performance (speed) with subjective assessments of satisfaction with the illumination level provided.

Bodmann's test task consisted of a card with the numbers 1 to 100 printed in a random manner over the whole area (Fig. 3.23 shows one of the charts, by courtesy of Prof. H. W. Bodmann). Beside each number was a small hole and the observer had to search the chart for the number called out by the experimenter and to touch through the adjacent hole a metal plate placed immediately under the card, using a thin metal pointer electrically connected to the recording equipment. When the experimenter called a number, he started at the same time a chronograph which was automatically stopped by contact between the pointer and the plate. This recorded the time taken by the observer to search for the number on the chart and to make the mechanical operation of touching the metal plate with the pointer. The size of the type face used for the numbers, and the contrast of the type with the background, was varied in the same way as in Weston's experiment. The observing session lasted for a period from about 10 to about 30 minutes. At the end of the session, the observer was asked to state whether the illumination level was 'too dark', 'good', or 'too bright'.

The results of the performance tests are shown in Fig. 3.24. Bodmann analysed his data for observers of 20 to 30 years of age (younger age group)

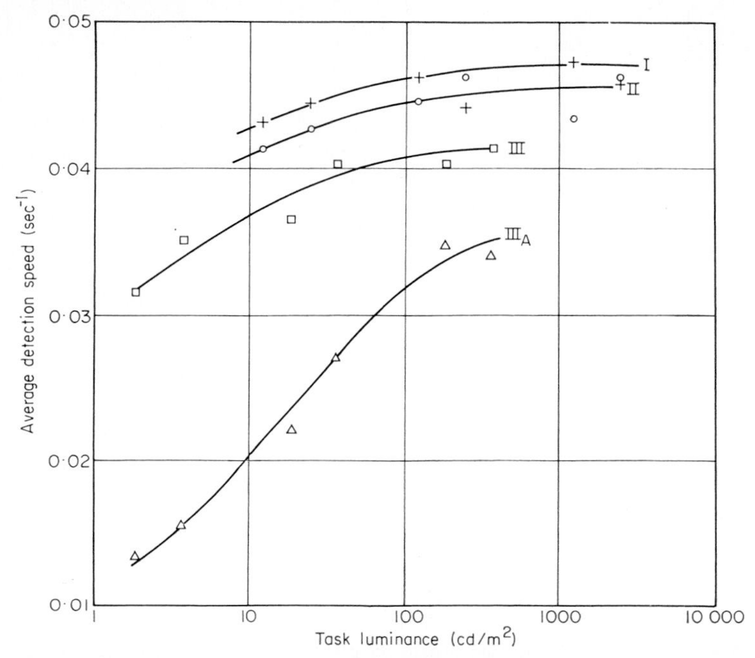

Fig. 3.24 *Relation between luminance level and visual performance achieved in a search task (printed numbers displayed at random) for the conditions indicated in Table below (From Bodmann 1967)*

Test	*Number of observers*	*Age range (years)*	*Back-ground reflection factor (%)*	*Object contrast*	*Angular size of detail (minutes)*	*Relative performance % (estimated from above graph)*
I	36	20–30	78	0·98	6	96
II	48	20–30	78	0·93	4	96
III	48	20–30	11·5	0·63	4	97
III_A	10	50–65	11·5	0·63	4	

and 50 to 65 (older age group). The data for the younger age group clearly showed a flattening of the curve relating speed of performance with illumination level when illumination was increased to give a luminance above 300 cd/m^2 to 3000 cd/m^2, i.e. with 1000 lux (100 lm/ft^2) to 10 000 lux (1000 lm/ft^2), the time taken being shorter at all values of illumination for the higher contrast tasks.

Bodmann's data also demonstrated that increased illumination assists older people in the performance of a visual task, and the older age group clearly benefited more from very high levels of illumination than did the younger age group, in that the corresponding curves do not show such a pronounced flattening. The data also showed a clearly defined subjective optimum level of illumination for each standard of task contrast. The maximum for the standard high contrast task is 1000 lux (100 lm/ft^2), and other optima apply to other tasks with the younger and the older age group (see Fig. 3.25).

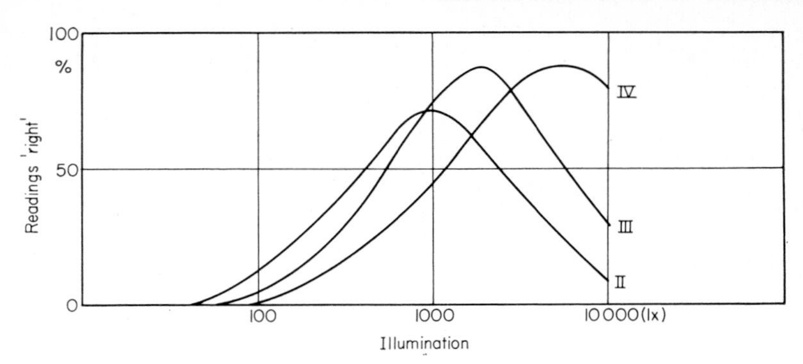

Fig. 3.25 *Percentage of subjective preferences against illumination* (*From Bodmann 1962*)
For locating numbers printed in black on white, most people, young as well as old, found a level of 1000 lux to be 'good' (II). For black figures on grey paper, most of the young people indicated 2000 lux as 'good' (III), while most of the older people favoured a level of 5000 lux (IV)

Bodmann's findings can be compared with the results obtained from Weston's formula and nomogram. The illumination levels obtained from Weston's formula give values quite near to maximum performance as determined by Bodmann. Weston's values are, however, much lower than the values chosen by Bodmann's observers subjectively.

Bodmann's experiment had some practical limitations which he pointed out. For example, the preferred values of illumination level may be limited in the upper direction by the fact that, at these very high levels, there may have been some discomfort from the radiant heat from his experimental lighting unit. He also suspects that the psychological conditioning of his observers to much lower levels of illumination may have influenced their judgment. It must also be borne in mind that the subjective assessment of levels of illumination must relate strictly to the type, colour and distribution of illumination provided, as well as to the distribution of the luminances in the visual field on the experimental arrangement. Consequently Bodmann's subjective preferences may not be representative, but it is of interest that his findings agree quite closely with rather similar work undertaken by Balder (1957).

Hopkinson (1965) has analysed the work of Bodmann and Balder and related it to other similar work, and came to the conclusion that there is adequate evidence for believing that there is an optimum level of illumination for given visual tasks, that increasing the illumination level beyond this optimum is of no value, and that in many cases it may be of distinct detriment in that glare discomfort from the excessive brightness of the task itself may be introduced.

This evidence seems to indicate that task luminance in the region of 80–200 ft-L (270–680 cd/m^2) (such as would be produced on white paper by an illumination of 1000–2000 lux or 100–200 lm/ft^2) is the optimum from the point of visual performance of a task involving readily visible detail in comfort and with minimum fatigue. This conclusion was reached by Simonson and Brozek in 1948, and has since been supported by a group of German workers (FORFA 1956), Kotova (1961) in Russia, Matsui and Kondo (1963) in Japan (for certain background conditions), Khek and Krivohlavy (1968) in Czechoslovakia and, more recently, by Boyce (1970) in this country.

Effect of Age on Visual Performance

All investigators—Weston (1949), Fortuin (1948), Bodmann (1962) and Weale (1963)—agree that older people see less well than younger people and derive more benefit from higher lighting levels.

Weale shows that as a result of miosis and lenticular yellowing, approximately one-third of the amount of light which reaches the retina of a 20-year old person can reach that of a person of 60, but according to Weston, an increase in illumination by this factor does not enable an old man to compare with a young man in visual performance, and that in fact it is impossible to compensate completely for the failing visual powers, by increasing illumination level, even if this is increased by some 1000 times.

Bodmann in a comparison between performances of a 20–30 age group and a 50–65 age group found that levels of performance (relatively low ones) achieved by the older subjects with illumination levels of 100–400 lux (10–40 lm/ft^2) were matched by the younger ones, in an illumination of only 2 to 5 lux (0·2–0·5 lm/ft^2).

It is therefore probably meaningless to indicate in a code of lighting a factor by which the illumination should be multiplied in relation to the age of those working under it. Such a factor will vary according to the levels of visual performance to be expected from the older workers.

These controlled laboratory tests under idealised conditions do not, however, necessarily indicate the true practical situation. Older workers are well known to be capable of very high degrees of skilled performance, and their handicap as judged subjectively by them is not always as great as the investigations would suggest. Some elderly people object to high levels of illumination which are tolerated by younger people. Factors which influence the overall situation include experience, manual and motor skills, the onset or delay of 'senile' changes in the eye, the original refractive errors of the eye, and others which indicate that lighting for older workers should be given individual consideration where possible.

Conclusions

The relation between lighting and visual performance has been studied by many investigators using different methods, but all come to more or less the same conclusion, that is, that increasing the illumination on a visual task up to a certain level will result in an improvement in visual performance, but that as this level is approached, the improvement in performance will become progressively less for the same increase in illumination level.

A normal-sighted person can achieve a good level of visual performance on normal high-contrast printing of 8-point type viewed at a normal reading distance with an illumination level of 100 lux (10 lm/ft^2). This same level of illumination enables high-contrast writing on a chalkboard to be read by normal-sighted persons without difficulty. Any decrease in illumination below this level, however, causes a rapid reduction in readability. For this reason, a value of 10 lm/ft^2 has been adopted, for example, by the Department of Education and Science as the absolute minimum to be installed in classrooms and teaching areas in schools.

The improvement to be registered in visual performance beyond this minimum level depends upon the nature of the task. The reading of high-contrast printing and writing is performed more easily when the illumination is increased up to about 1000 lux (100 lm/ft^2) but beyond this the benefits are minimal and problems may arise from glare discomfort from the brightness of the task itself. On the other hand, other more difficult tasks involving smaller critical detail and lower contrast benefit from increased levels of illumination to a very much greater degree. British and North American practice, based respectively on work of Weston and Blackwell, proposes ranges of illumination levels related to the visual difficulty of the task. Although the illumination levels recommended by each are different, nevertheless (because of the flattened nature of the performance-illumination curve) these differences are of little scientific significance though obviously of great economic import, and it is on the latter criterion in relation to 'status' or attractiveness of the work place that the choice must be made from within the range permitting adequate performance of the task in question.

References

Balder, J. J. (1957): 'Erwünschte Leuchtdichten in Büroräumen. *Lichttechnik*, **9**, 455–461.

Beuttell, A. W. (1934): 'An Analytical Basis for a Lighting Code'. *Illum. Engr.* (*London*), **27**, 5–16.

Blackwell, H. R. (1946). 'Contrast Thresholds of the Human Eye.' *J. Opt. Soc. Amer.*, **36**, 624.

Blackwell, H. R. (1959): 'Specification of Interior Illumination Levels on the Basis of Performance Data.' *Illum. Engng.* (*New York*), **54**, 317–353.

Bodmann, H. W. (1961) (with E. Muck): 'Die Bedeutung des Beleuchtungsniveaus bei Praktischer Sehtätigkeit.' *Lichttechnik*, **10**, 502–507.

Bodmann, H. W. (1962): 'Illumination Levels and Visual Performance.' *Internat. Lighting Rev.*, **13**, 41–47.

Bodmann, H. W. (1967): 'Quality of Interior Lighting Based on Luminance.' *Trans. Illum. Engng. Soc.* (*London*), **32**, 22–40.

Boyce, P. R. (1970). 'The Influence of Illumination Level on Prolonged Work Performance'. *Lighting Research and Technology* (in press).

CIE (1968): 'Report of Committee E. 1.4.2. on Visual Performance.' Proc. CIE 16th Session (Washington, 1967). Commission Internationale de l'Éclairage, Paris.

Dunbar, C. (1940): 'Fundamental Principles of Meters used to Measure Visibility'. *Trans. Illum. Engng. Soc.* (*London*), **5**, 33–43.

FORFA (1956): 'Investigation into the Efficiency and Fatigue of the Human Being under Various Conditions of Illumination.' *Lichttechnik*, **8**, 296–300. (In German. Translation in BRS Library Communication No. 821, 1958.)

Fortuin, G. J. (1948). 'The Effect of Illumination and Age on Visual Acuity.' Proc. Ninth Internat. Congress on Industrial Medicine, London.

Foxell, C. A. P. and W. R. Stevens (1955). 'Measurements of Visual Acuity'. *Brit J. Ophthalmol.*/ **39**, 513–533.

Fry, G. A. and J. M. Enoch (1959): 'Visual Perception of the Orientation of a Landolt Ring.' MCRL Technical Paper No. (696)–19–292. Ohio State University, Columbus, Ohio.

Fry, G. A. (1962). 'Assessment of Visual Performance.' *Illum. Engng.* (*New York*), **57**, 426–437.

Gilbert, M. and R. G. Hopkinson (1949): 'The Illumination of the Snellen Chart.' *Brit. J. Ophthalmol.*, **33**, 305–310.

Guth, S. K., A. A. Eastman and R. C. Rodgers (1953): 'Brightness Differences—A Basic Factor in Suprathreshold Seeing.' *Illum. Engng.* (*New York*), **48**, 233–239.

Hopkinson, R. G. (1946): 'Visibility of Cathode Ray Tube Traces.' *J. Instn. Elect. Engrs.* **93**, (111A), 795–807.

Hopkinson, R. G. (1949): 'Studies of Lighting and Vision in Schools.' *Trans. Illum. Engng. Soc.* (*London*), **14**, 244–268.

Hopkinson, R. G. (1951): 'The Brightness of the Environment and its Influence on Visual Comfort and Efficiency.' Proc. Building Research Congress, Div. 3, Part III, 133–138. HMSO, London.

Hopkinson, R. G. (1965): 'A Proposed Luminance Basis for a Lighting Code.' *Trans. Illum. Engng. Soc.* (*London*), **30**, 63–88.

IES, New York (1966): IES Handbook, 4th Edition, pp. 9–49 to 9–63. Illuminating Engineering Society, New York.

Khek, J. and J. Krivohlavy (1968): 'Evaluation of the Criteria to Measure the Suitability of Visual Conditions.' Proc. CIE 16th Session (Washington, 1967). Paper P67.19. Commission Internationale de l'Éclairage, Paris.

Kotova, E. L. (1961): 'Optimum Level of Illumination for Fine Visual Work.' *Svetotekhnika*, **7**, 19–24. (In Russian.)

Luckiesh, M. and F. K. Moss (1934): 'A Visual Thresholdometer.' *J. Opt. Soc. Amer.*, **24**, 305–307.

Lythgoe, R. J. (1932): 'The Measurement of Visual Acuity.' Medical Research Council Special Report No. 173.

Matsui, M. and M. Konda (1963): 'Studies on Relation Between Illumination Levels and Visual Fatigue.' *J. Illum. Engng. Inst. Japan*, **47**, 176–186. (In Japanese with English summary.)

Simonson, E. and Brozek, J. (1948): 'Effects of Illumination Level on Visual Performance and Fatigue.' *J. Opt. Soc. Amer.*, **38**, 384–397.

Stevens, W. R. and C. A. P. Foxell (1955): 'Visual Acuity.' *Light and Lighting*, **48**, 419–424.

Stevens, W. R. and C. A. P. Foxell (1955): 'Visual Acuity.' *Brit. J. Ophthalmol.*, **39**, 513.

Weale, R. A. (1963). 'The Aging Eye.' H. K. Lewis, London.

Weston, H. C. (1945): 'The Relation between Illumination and Visual Performance.' Industrial Health Research Board Report No. 87 (reprinted 1953). HMSO, London.

Weston, H. C. (1949): 'On Age and Illumination in Relation to Visual Performance.' *Trans. Illum. Engng. Soc.* (*London*), **14**, 281–297.

Weston, H. C. (1961): 'Rationally Recommended Illumination Levels.' *Trans. Illum. Engng. Soc.* (*London*), **26**, 1–16.

Weston, H. C. (1962): 'Light, Sight and Work.' H. K. Lewis, London.

Weston, H. C., D. J. Bridgers and J. Ledger (1963): 'A Study of the Effect of Pattern on the Detection of Detail at Different Levels of Illumination.' *Ergonomics*, **6**, 367–377.

4

Glare Evaluation and Control

The Evaluation of Glare

When there are areas of very high brightness in the visual field, some form of glare will result. If there is direct interference with vision, the condition is called 'disability glare' (see Chapter 1); if this disability persists after the immediate cause is removed, the condition is called 'successive glare' and is due to the time necessary for the processes of adaptation to readjust. If there is no direct interference with vision, but discomfort, or annoyance, irritation or distraction, the condition is called 'discomfort glare'. In practice all these conditions may arise from the same glare cause, but in interior lighting to high or moderate levels, the usual cause of complaint is the discomfort rather than the disability which the presence of bright light sources introduces to the visual field.

Disability Glare

The origins of glare which causes disability to vision are at least twofold, the one a veil of scattered light in the optic media which reduces contrast perception, and the other an inhibitory effect in the retinal processes which results in a diminution of the apparent brightness of the object of regard. This inhibitory effect in turn is due partly to photochemical and partly to electrical action in the retina and neural pathways.

Luckiesh and Holladay (1925), Holladay (1926) and later Stiles and his colleagues (1928 *et seq.*) and Crawford and Stiles (1935, 1937) showed that the disabling effects of a glaring source in the visual field could be quantified in terms of an 'equivalent veiling luminance' over the field. Thus if a target of 2 units of luminance can just be seen in a field of 100 units (i.e. a 'contrast threshold' of 2% on the ratio basis of assessing contrast—see Chapter 3), the effect of an equivalent veiling luminance of 10 units due to a glare source is to add this veil to both target and background field, so that the contrast now becomes $(2 + 10)/(100 + 10) = 1{\cdot}09\%$. Since the threshold value is 2%, the contrast will no longer be visible, and so the effect of the glare source will have been to render something previously visible no longer so. Disability glare on a street at night has precisely this result, whether due to an oncoming car headlight, or, less likely, due to glare from improperly shielded street lighting units.

Holladay showed that this equivalent veiling brightness could be expressed by the formula:

$$L_v = kE/\theta^2 \tag{4.1}$$

where E is the illumination produced by the glaring source at the eye, and θ is the angular distance from the line of sight to the direction of the glare source. The work of Stiles and of others who followed confirmed the findings of Holladay. More recently Christie and Fisher (1966) have shown that the value of the constant k in Holladay's formula can be related to the age of the observer in a manner which would be explained by the loss in clarity of the optic media with increasing age.

There are two important features of Holladay's findings which are of practical significance: first, that the greater the angular distance of the glaring source from the direction of viewing, the less the veiling effect; and second, that the effect is the same whether the glaring source is a small source of high luminance or a large source of low luminance, provided that the illumination on the eye (E) is the same. Thus a large window backed by a low-brightness sky can cause marked disability glare and reduce contrasts of neighbouring objects, perhaps below the threshold of visibility, even though its brightness may not be sufficient to cause any discomfort.

While disability glare is of importance on lighted streets at night, and generally in all situations where the amount of available light renders some objects of significance at or near the threshold of vision, it is an effect of only minor significance in modern interior lighting practice. The problem of glare in interior lighting is therefore primarily one of discomfort caused by excessively bright areas in the visual field. This excessive brightness may result from direct view of bare lamps in badly designed fittings, from a large area of an excessively bright luminous ceiling, or from an excessively bright sky seen through a window. In artificial lighting design the control of glare is the designer's highest priority after providing the necessary working illumination for visual efficiency. Under lighting conditions characteristic of current practice, the levels of working illumination which are provided are usually fully adequate for the satisfactory performance of the visual task, but still much remains to be done to eliminate glare which, in its milder form, may cause no more than irritation or distraction but which unfortunately often causes real discomfort.

Discomfort Glare

The discomfort caused by glare appears to have entirely different physiological origins from the disabling effect. These origins have not been studied in the same detail as have the effects of direct disability (Holladay, Stiles,

Schouten, Ivanoff, Christie and Fisher). Fugate and Fry (1956) have shown that the sensation of glare discomfort is linked to the activity of the iridomotor muscles which control the pupil diameter, but Hopkinson (1956) showed that while this link between the iridomotor system and the sensation of glare discomfort existed, it was by no means a direct link, situations being both possible and common in which a negative relation between pupil diameter and glare sensation existed. Hopkinson concluded that the sensation of discomfort was associated, at least in part, with opposing actions of sphincter and dilator muscles, due to an instability arising from contradictory indications from the highly stimulated parts of the retina, receiving illumination from the glare sources, and from the less fully stimulated parts of the retina, receiving illumination from a low-luminance surround.

Even so, all that these authors accomplished was to show that the sensation of discomfort might be related to iridomotor activity, but they did not demonstrate how this activity comes about. The activity must clearly be retinal in origin, presumably due to 'neural leakage' between the high activity of the neural pathways from the retina to the brain of those elements stimulated by the high-luminance glare source, and the neighbouring pathways of the elements stimulated by the low-luminance background. The exact nature of this neural leakage has not been established, nor is it known at which stage of the transmission line to the higher visual centres of the brain the information is passed to the iridomotor system.

The study of glare discomfort has therefore been entirely empirical, but this has not prevented sufficient information being obtained to enable any glare situation to be evaluated quantitatively, and for quantitative measures to be prescribed for the control or elimination of glare discomfort. Success in this field, however, awaited the design of experiments which permitted each of the variables of the visual field (the source, its brightness and size, its position in the visual field, and the background against which it was seen) to be studied separately, with all others controlled at fixed values. Sociological surveys had only revealed the broad nature of the glare phenomenon without giving any precise numerical assistance to the designer. The investigation of glare is, in fact, a classic case of the severe limitations of sociological methods for the quantitative evaluation of aspects of environmental design.

Experimental Methods of Quantifying Glare Discomfort Effects

The basic studies on glare discomfort have employed forms of apparatus in which the glare source can be varied in size and brightness, while the rest of the visual field remains exactly the same, although apparently lit by the source. One of the methods employed was that of Hopkinson (1940 *et seq.*), Hopkinson and Petherbridge (1954), and Petherbridge and Hopkinson (1950), in which the observer looks into a model room in which there are apertures corresponding to the glare sources under investigation. The luminance of these apertures is provided by an illuminating system outside the model itself.

The luminance of the interior of the model is provided by lamps (hidden from the observer) under separate control. Various forms of optical system were used in the different investigations to ensure that any change in the luminance of the glare sources would not result in any effect on the luminance of the general surroundings and vice versa.

The method of evaluation of glare discomfort has been to ask observers for their direct subjective impressions. In the earliest work (Luckiesh and Holladay, 1925) the procedure was to ask the observer to alter the luminance of the glare source to give in turn a series of sensations of discomfort or comfort. A very large number of sensations from 'painful' to 'very pleasant' were used, which gave rise to criticism as being meaningless. Hopkinson (1940) developed a method which employed a series of not more than four precisely described borderline criteria of glare discomfort ('just perceptible', 'just acceptable', 'just uncomfortable', 'just intolerable'), associating with these criteria a method of 'calibrating' the human observer over a long period so that his evaluations could be improved in precision by experience, and the variance of these evaluations about a mean could be determined. Luckiesh and Guth (1949) used a single criterion of discomfort called the 'borderline between comfort and discomfort' or 'BCD', rather than a number of criteria.

Investigations on glare discomfort have therefore used 'introspective' methods of evaluation, that is, the observer is asked to think about the situation, evaluate it, and make a setting of a controlled variable, such as the source luminance, until the glare sensation corresponds to a criterion which has been described to him and which he believes he can reproduce. No successful studies have been made using specific visual tasks and the performance of them as the basis for evaluation. Attempts have indeed been made to evaluate glare in this way (Bartlett and Pollock 1935, Stone and Groves 1968) but all the investigators who have attempted such 'behaviourist' investigations have found that acute subjective discomfort arises from situations which give rise to little or no decrement in visual performance. If such decrement does exist, it probably only results from very long exposures, that is, over days, weeks or months to a glaring situation. Glare is essentially a matter of annoyance and distress rather than direct reduction in visual performance.

It is necessary to mention here the difference in approach of one group of American investigators (Luckiesh 1925 *et seq.*) as compared with all others. Luckiesh, and Guth, later his colleague, have insisted, from the earliest days of Luckiesh's work, on presenting the glare source momentarily rather than continuously. Luckiesh's argument was that glare is experienced when the observer looks up from his work and sees the glare source for a second or so and then looks down at his work again. Luckiesh therefore designed his experiments so that the glare source would be exposed in a sequence (a ten-second cycle, with three one-second 'on' periods separated by one-second 'off' periods, followed by a five-second 'off' period before the next 'on' period). Other investigators have used continuous exposure of the glare source, on the hypothesis that in a normal interior, the glare sources like any other part of the field of view are continuously exposed and have a continuous influence.

The results of the various investigations into the glare discomfort phenomenon all agree that the magnitude of glare sensation is related directly to the luminance of the glaring source and its apparent size as seen by the observer, and that the discomfort is reduced if the source is seen in surroundings of high luminance. The glare sensation is also reduced the further the glare source is off the line of sight. Over a limited range of conditions, this finding can be expressed by a formula as follows:

$$\text{Glare constant} = (B_s^{1\cdot 6}\omega^{0\cdot 8})/B_b\theta^2 \qquad (4.2)$$

where B_s is the luminance of the source,
ω is the solid angle subtended by the source,
B_b is the general background luminance,
θ is the angle between the direction of viewing and the direction of the glare source.

The exponents in the above formula are themselves dependent upon the values of the parameters, and are therefore not constants. It is usual in the above 'glare formula' to substitute for the θ-term a factor called a 'position index' p_g which is an expression of the complicated way in which the position effect varies with the vertical and horizontal displacement of the glare source from the direction of viewing.

One of the consequences of the different experimental techniques used by Luckiesh and Guth on the one hand (momentary exposure) and by other investigators on the other hand (continuous exposure) is that the exponent of the glare source in the above formula is markedly different, the momentary exposure technique resulting in a higher influence of the luminance of the glare source (and therefore a higher exponent of the B_s term in the above formula) (Hopkinson 1957). This basic difference in procedure and results has held up international agreement on the evaluation of glare.

Such a formula enables the magnitude of the glare sensation from a single source to be evaluated. The glare constant obtained in this way can be related directly to the sensation of glare. For example, if in the above formula (4.2) the luminance of the source and surround are expressed in foot-lamberts, the apparent area of the source in steradians, and the 'position effect' suitably expressed as a pure ratio, the values of glare constant derived from the formula are found to correspond as follows (Table 4.1):

Table 4.1

Glare Constant	*Criterion of Glare Discomfort*
600	Just intolerable
150	Just uncomfortable
35	Just acceptable
8	Just perceptible

Thus, if the glare constant for a given situation was evaluated as 60, the degree of glare, for the average observer, would be between 'just uncomfortable' and 'just acceptable'.

Glare from a Number of Sources

The simple glare formula applies to a single source, but it has been shown that the effect of a number of sources can be obtained by suitable addition of the constants obtained for each individual source in the field of view. Hopkinson (1940) showed that the simple arithmetic addition of the glare constants gave a value of glare constant which corresponded closely to the sensation from the complete array of multiple sources. Petherbridge and Hopkinson (1950) modified this proposal by showing that the glare from a number of sources is the same as that from a single source of equal apparent area placed at the centroid of the array. Later, Hopkinson (1957) showed, however, that the additive nature of glare was a highly complex phenomenon, the exact additivity function depending upon the luminance of the sources and their position in the field of view, among other things. A detailed experimental investigation of the additivity function predicted the conditions in which simple arithmetic additivity of the glare constants would be sufficiently accurate for practical purposes, and the conditions in which it would not, together with an indication of the magnitude and sign of the errors so involved. Other workers, notably Einhorn (1961), have put forward alternative arithmetical proposals for the additivity of glare constants. Guth (1963) has proposed a method of adding the glaring effects of single sources which involves an exponential additivity function, the exponent being related to the number of sources. In Great Britain the simple arithmetic additivity of glare constants has been incorporated by the Illuminating Engineering Society in its method of glare prediction (Technical Report No. 10, 1967), but the limitations inherent in the procedure are pointed out.

Variance of Glare Evaluations

The evaluation of glare discomfort depends upon the subjective assessments of observers who are in experimental control of the glare situation. They have to adjust one of the possible variables (e.g. source luminance) to give, in Hopkinson's experiments, a described series of criteria of discomfort or, in Guth's experiments, a single criterion, the borderline between comfort and discomfort (BCD). The variance of settings made in this way over a long period of time has been established. For example, Hopkinson (1955) showed that observers who have acquired some experience in making such observations of glare discomfort maintain their criteria over a very long period of time (several years) and the variance about the mean remains more or less steady for a given observer. He suggests that the variability of an observer is about $\pm 0{\cdot}15$ on a logarithmic scale of luminance. On the other hand, inexperienced observers tend to change their sensitivity, and therefore their assessments, as they gain experience. Usually this change of sensitivity has an upward trend, that is, observers become more sensitive to glare as the experiment progresses, eventually reaching a steady state, about which mean the variance of their observations can be computed.

Once a steady state has been reached, however, individual observers do not give the same settings. Some are demonstrably more sensitive to glare than others, and this difference in sensitivity is maintained throughout a long series of experiments. These differences may be physiological in origin or they may equally be due to some psychological difference, in that some people may be more tolerant of a discomfort which to them is as great as it might be to another, less tolerant, individual. No systematic work has been done to establish whether these differences in individual tolerance are associated with, for example, sex, age, or systematic visual defect. It is, however, of interest that in the course of many years' experimentation, a few subjects with high negative refractive error have all been found to be significantly

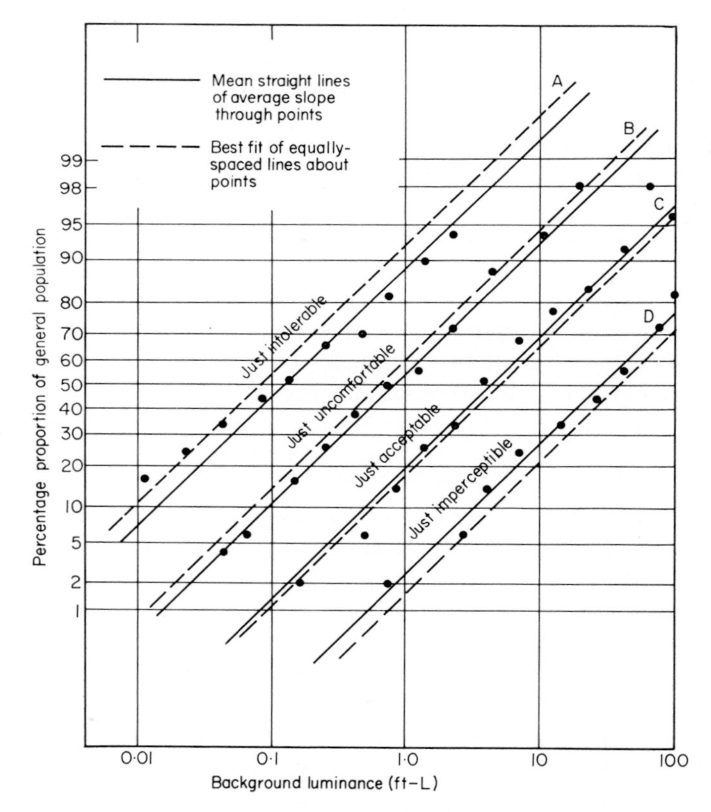

Fig. 4.1 *Variation with background luminance of the percentage proportion of the general population who would assess glare discomfort in a particular environment at the given level or less uncomfortable than the given level (Source luminance = 3000 ft–L)*

less sensitive to glare than the general population, and this has been confirmed anecdotally.

The variance in the sensitivity of the general population can be expressed by means of a probability function as shown in Fig. 4.1. The variance as shown in this way is very great, because the data was obtained by asking a large number of individuals to make assessments for the first time. With experience and skill their judgments would be refined and would come closer to a common mean. In practice, lighting codes are not based upon the average of the assessments of such a group of unskilled observers, but are set at the 85% or 90% level on the probability function in order to ensure that the majority of the general population will find the glare situation at least acceptable if not imperceptible.

The Control of Glare in Lighting Design

It is evident that the existence of an empirical formula for the evaluation of glare sensation can be used for the design of glare-free lighting. Provided that the sum of the glare constants from an array of lighting units in a room is less than a total of 8, it follows from Table 4.1 that glare will then be imperceptible. In the same way, if the sum of the glare constants is greater than 600, glare will then be intolerable. These figures, of course, apply to the population for whom the correlation given in the table can be taken to apply.

The system which has been devised in Britain for the control of glare in artificial lighting installations depends upon this property of the empirical glare formula. The Illuminating Engineering Society of Great Britain (Techical Report No. 10, 1967) has devised a method of evaluating the sum of the glare constants for an array of lighting units in a room; this sum is called a 'Glare Index'. However, instead of relating this Glare Index directly to the sensations of glare as has been done in Table 4.1, the procedure instead is to set limiting values of this Glare Index in such a way that experience indicates that complaints from glare discomfort will be few or entirely absent. This would, of course, apply if the Glare Index was always set to a value corresponding to less than 8 in the table. However, it was recognised that in some situations tolerance of glare would be much greater than in others. In a factory store-room, where people may go in only for brief periods, intent on a particular activity, looking for a piece of material or equipment, the degree of glare which might cause complaint is probably far higher than that in a hospital ward, where people are lying, sometimes sick, in such a way that they may not be able to turn away their gaze from the direction of a glaring light source. The limits on the Glare Index have therefore been set in the IES Code (1968) as a result of field studies in a wide range of environments still leaving it open to modify the values of the Glare Index limits in the IES Code, should experience demand it.

It is evident that the calculation of a Glare Index using the empirical formula would necessarily be lengthy in order to evaluate the glare constant for

Table 4.2

INITIAL GLARE INDEX: LIGHT DISTRIBUTION CLASSIFICATION BZ 5

Flux fraction ratios of lighting fittings (upper/lower)

Reflection factors of room surfaces (per cent)

Room dimension		$0\left(\frac{UFF}{LFF}=\frac{0\%}{100\%}\right)$					$0{\cdot}33\left(\frac{UFF}{LFF}=\frac{25\%}{75\%}\right)$					$1{\cdot}0\left(\frac{UFF}{LFF}=\frac{50\%}{50\%}\right)$					$3{\cdot}0\left(\frac{UFF}{LFF}=\frac{75\%}{25\%}\right)$				
	Ceiling	70	70	50	50	30	70	70	50	50	30	70	70	50	50	30	70	70	50	50	30
	Walls	50	30	50	30	30	50	30	50	30	30	50	30	50	30	30	50	30	50	30	30
	Floor	14	14	14	14	14	14	14	14	14	14	14	14	14	14	14	14	14	14	14	14
X	*Y*	*Initial Glare Indices*																			
2*H*	2*H*	18·0	20·4	18·4	20·9	21·3	15·7	17·7	16·5	18·6	19·7	13·2	14·8	14·4	16·0	17·7	9·6	10·8	11·0	12·2	14·2
	3*H*	20·7	23·0	21·1	23·3	23·7	18·4	20·2	19·2	21·0	22·1	15·7	17·1	16·8	18·3	20·1	12·0	13·3	13·4	14·7	16·6
	4*H*	21·9	24·1	22·4	24·5	24·8	19·5	21·1	20·5	22·0	23·1	16·9	18·1	18·1	19·3	20·9	13·1	14·1	14·5	15·6	17·5
	6*H*	22·9	25·1	23·4	25·4	25·8	20·5	22·1	21·4	23·0	24·1	17·7	18·9	18·9	20·2	21·8	14·1	15·0	15·5	16·5	18·5
	8*H*	23·2	25·3	23·8	25·8	26·2	20·9	22·3	21·8	23·3	24·4	18·2	19·3	19·5	20·6	22·2	14·4	15·4	15·9	16·9	18·8
	12*H*	23·9	25·9	24·5	26·4	26·7	21·6	22·8	22·5	23·9	25·0	18·6	19·7	19·8	21·0	22·7	14·9	15·7	16·3	17·2	19·2
4*H*	2*H*	19·2	21·5	19·7	21·8	22·2	16·9	18·4	17·8	19·3	20·4	14·2	15·4	15·4	16·6	18·3	10·4	11·4	11·9	12·9	14·9
	3*H*	22·2	24·2	22·7	24·6	25·0	19·8	21·1	20·8	22·2	23·3	16·9	18·0	18·1	19·3	21·0	13·1	13·9	14·6	15·5	17·5
	4*H*	23·8	25·6	24·3	26·0	26·5	21·2	22·3	22·2	23·4	24·6	18·5	19·3	19·8	20·6	22·3	14·4	15·0	15·9	16·7	18·5
	6*H*	24·7	26·5	25·3	26·9	27·5	22·1	23·3	23·1	24·3	25·5	19·3	20·0	20·5	21·5	23·2	15·4	16·0	16·9	17·6	19·5
	8*H*	25·5	26·9	26·0	27·4	28·0	22·8	23·7	23·7	24·8	26·0	19·9	20·7	21·2	22·1	23·7	16·0	16·5	17·5	18·1	19·9
	12*H*	26·0	27·5	26·5	27·9	28·5	23·3	24·2	24·3	25·3	26·5	20·5	21·1	21·7	22·5	24·1	16·6	17·0	18·1	18·7	20·5

Table 4.2 (*contd.*)

8*H*	4*H*	24·4	25·9	24·9	26·3	26·9	21·7	22·6	22·7	23·7	24·9	18·9	19·6	20·1	21·0	22·6	14·8	15·5	16·3	17·1	18·9
	6*H*	26·1	27·3	26·7	27·8	28·5	23·4	24·1	24·4	25·1	26·5	20·4	20·9	21·7	22·2	23·9	16·4	16·8	17·8	18·5	20·4
	8*H*	26·8	28·0	27·5	28·6	29·2	24·0	24·8	25·1	26·0	27·2	21·0	21·6	22·4	23·0	24·6	17·1	17·5	18·6	19·2	21·1
	12*H*	27·3	28·5	28·0	29·1	29·7	24·6	25·3	25·6	26·5	27·7	21·7	22·2	23·0	23·7	25·3	17·7	18·0	19·2	19·7	21·6
12*H*	4*H*	24·6	26·1	25·1	26·5	27·1	21·9	22·8	22·9	23·9	25·1	19·1	19·7	20·3	21·1	22·7	15·2	15·7	16·6	17·3	19·1
	6*H*	26·3	27·5	27·0	28·1	28·7	23·6	24·3	24·7	25·5	26·7	20·6	21·1	21·9	22·6	24·2	16·7	17·1	18·2	18·8	20·6
	8*H*	27·0	28·2	27·7	28·7	29·4	24·2	25·0	25·3	26·2	27·4	21·3	21·9	22·7	23·4	25·0	17·4	17·7	18·8	19·4	21·3
	12*H*	27·5	28·8	28·3	29·4	30·0	24·8	25·6	26·0	26·6	28·0	22·0	22·6	23·4	23·9	25·7	18·1	18·3	19·5	19·9	22·0

H Height of fitting above 4 ft eye level.
X Room dimension at right angles to the line of sight in terms of the height *H*.
Y Room dimension parallel to the line of sight in terms of the height *H*.

all the light sources in a particular installation of a large number of sources. The IES therefore produced tables for installations which consist of regular symmetrical arrangements of lighting fittings with downward light distributions of ten different standard types in rooms of various dimensions and with surfaces of various reflection factors. These tabulated values of Glare Index in IES Technical Report No. 10 are the basis of the IES Glare Index system. With the aid of this system, the Glare Index can be found rapidly from the known photometric and geometric constants of the installation.

The procedure for the evaluation of the Glare Index on the IES system is briefly as follows: Reference should be made to Tables 4.2 and 4.3, which are reproduced by permission of the Illuminating Engineering Society for the purpose of illustrating the operation of the system. Table 4.2 is one of a series of ten tables which give the Initial Glare Index for a particular form of light distribution, in this case with a distribution corresponding to a 'perfect diffuser'. There are corresponding tables for ten classifications of light distribution corresponding to the ten distributions in the British Zonal System (BZ 1 to BZ 10) (see Chapter 2 and Fig. 4.2). Table 4.3 is a conversion table by which the Initial Glare Index obtained from Table 4.2 is modified according to the downward light output, the luminous area of the lighting fitting, and the mounting height of the lighting units above the eye level.

In order to determine the Glare Index for a complete installation of symmetrically arranged lighting fittings in a room, it is necessary to know:

1. The room dimensions* as a function (X and Y) of the height of the fittings above a 4 ft (1·2 m) eye level. Thus a room 20 ft × 16 ft (6 m × 4·8 m) with the fittings mounted 8 ft (1·2 m) above the floor would have the dimensions $X = 5$, $Y = 4$.
2. The downward light distribution from the fittings in terms of the nearest classification in the British Zonal (BZ) system. This information would normally be available from the manufacturer. If not, the polar curve of light distribution would have be to measured, which would then be compared with the BZ classification (Fig. 4.2) and the appropriate number in the BZ system selected. (This comparison can be made visually if the downward light distribution is reasonably smooth. If not, the zonal flux relationships may have to be compared.)
3. The flux fraction ratio (FFR), that is, the ratio of the total light flux from the fitting which is received above the horizontal to that which is received below the horizontal, a figure which would normally be supplied by the manufacturer of the lighting equipment, but which would otherwise have to be computed from the measured complete polar curve of light distribution.
4. The total downward flux F in lumens.
5. The average reflectances of the ceiling, walls and floor (these figures can

* The introduction of metric or SI units does not involve any change in the tables of Initial Glare Index.

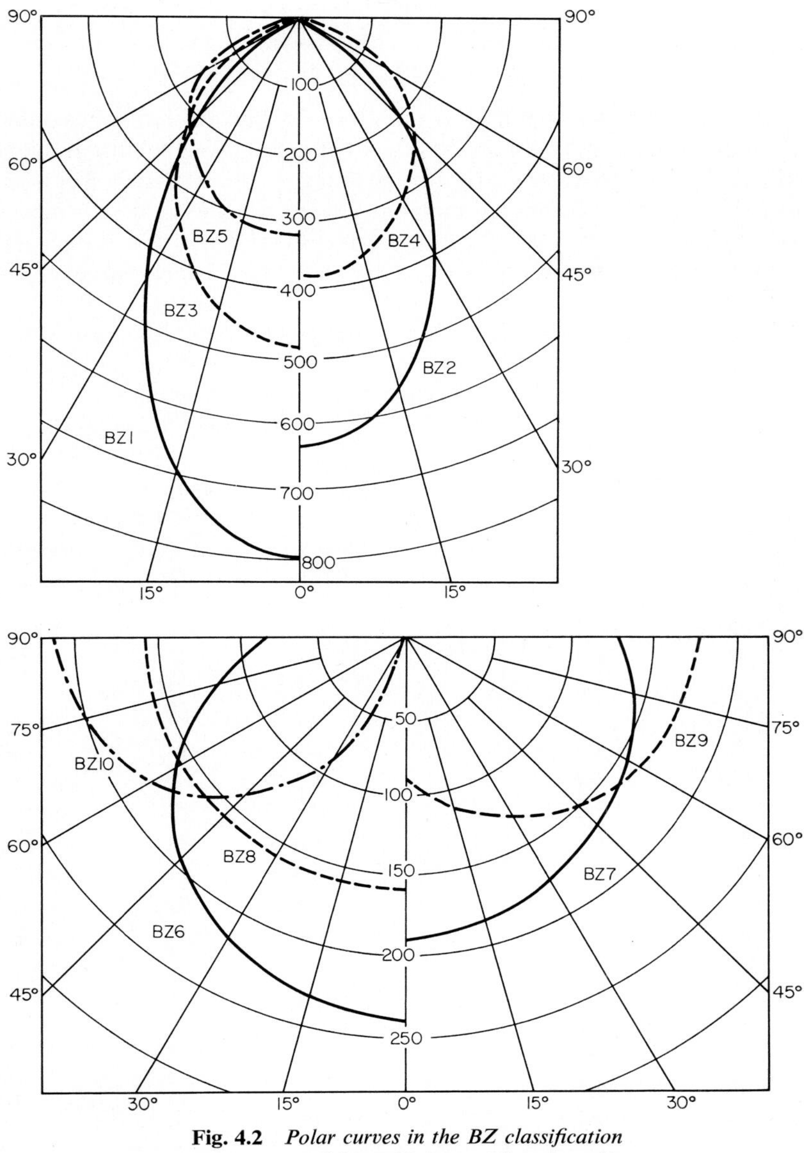

Fig. 4.2 *Polar curves in the BZ classification*

BZ 1	$I \propto \cos^4 \theta$	BZ 6	$I \propto (1 + 2 \cos \theta)$
BZ 2	$I \propto \cos^3 \theta$	BZ 7	$I \propto (2 + \cos \theta)$
BZ 3	$I \propto \cos^2 \theta$	BZ 8	I constant
BZ 4	$I \propto \cos^{1.5} \theta$	BZ 9	$I \propto (1 + \sin \theta)$
BZ 5	$I \propto \cos \theta$	BZ 10	$I \propto \sin \theta$

The BZ classification relates to the lower hemisphere only; the polar curves above are scaled to give 1000 lumens in the lower hemisphere for purposes of comparison

Table 4.3

GLARE INDEX CONVERSION TERMS FOR DOWNWARD FLUX, LUMINOUS AREA AND HEIGHT ABOVE 4 FT EYE LEVEL

Conversion terms corresponding to the values of downward flux *F*, luminous area *A* and mounting height *H* above a 4 ft eye level for the fittings actually used are obtained from the Table interpolating where necessary. These three conversion terms are added algebraically, taking account of the positive and negative signs, and the sum (which may be positive or negative) is then added to or subtracted from the Initial Glare Index for the installation taken from Table 4.2.
The downward flux *F* is the total flux output per fitting in lumens multiplied by the lower flux fraction.
The area *A* is the luminous area in square inches of each fitting.
The height *H* is the height in feet of the fittings above a 4 ft eye level.

Downward flux F (lm)	*Conversion term*	*Luminous area A (in^2)*	*Conversion term*	*Height H above 4 ft eye level (ft)*	*Conversion term*
100	−6·0	10	+8·0	3	−1·3
150	−4·9	15	+6·6	4	−1·0
200	−4·2	20	+5·6	6	−0·6
300	−3·1	30	+4·2	8	−0·3
500	−1·8	50	+2·4	10	0·0
700	−0·9	70	+1·2	12	+0·3
1000	0·0	100	0·0	15	+0·6
1500	+1·1	150	−1·4	20	+1·0
2000	+1·8	200	−2·4	25	+1·3
3000	+2·9	300	−3·8	30	+1·6
5000	+4·2	500	−5·6	40	+2·1
7000	+5·1	700	−6·8		
10 000	+6·0	1000	−8·0		
15 000	+7·1	1500	−9·4		
20 000	+7·8	2000	−10·4		
30 000	+8·9	3000	−11·8		
50 000	+10·2	5000	−13·6		

The data on which the IES Glare Index System is based are restricted at present to sources which have a maximum solid angle subtense at the eye of 0·1 steradian. Therefore, for the larger luminous areas quoted here, while the system is applicable when they are used at high mounting, it cannot strictly be employed for them at low mounting, but the errors involved are likely to be small.

be approximated by simple arithmetical evaluation of individual areas and reflectances).

6. The luminous area* of the fitting. For the purposes of the calculations this is taken as the downward projected area for BZ classifications of 1 to 8, and the horizontally projected area for BZ classifications of 9 and 10.

The first step is to find the Initial Glare Index from the table similar to Table 4.2 for the appropriate BZ classification. The appropriate column for the flux fraction ratio is selected, and the room dimensions X and Y are found and related to the appropriate values of the room surface reflectances. Thus, for example, if the flux fraction ratio is 1 : 3 (25% upwards, 75% downwards), and the reflectances of the ceiling, walls and floor are respectively 73%, 28% and 17%, the seventh column from the left of Table 4.2 (corresponding to ceiling, walls, floor 70: 30 : 14) would be selected as the nearest to the actual conditions. For room dimensions of 20 ft × 16 ft (6 m × 4·8 m) for a mounting height H of 4 ft (1·2 m) above a 4 ft (1·2 m) eye level, the horizontal line for $X = 4H$, $Y = 4H$ would give a value of Initial Glare Index of 20·3, while the horizontal line for $X = 4H$, $Y = 6H$ would give a value of 20·5. The actual room whose dimensions are $X = 4H$, $Y = 5H$ would therefore by direct linear interpolation have an Initial Glare Index of 20·4.

The next stage would be to correct this Initial Glare Index for the actual values of downward flux F, for the actual luminous area A, and for the height H above a 4 ft eye level, using Table 4.3. (The tables in the Appendix to Technical Report No. 10 would be used in conjunction with metric dimensions.) Suppose the downward flux to be 3000 lumens, the luminous area of the fittings to be 200 in^2; it can be seen from the tables that the conversion term for downward flux is +2·9, for luminous area is −2·4, while the conversion factor for the actual height H above the 4 ft eye level (in this case 4 ft) is −1·0. The sum of these conversion factors is therefore 2·9 − 2·4 − 1·0 = −0·5. These conversion factors are arithmetic; the value of Glare Index for the installation is therefore 20·4 − 0·5 = 19·9.

Further correction factors may also be applied, but the procedure is equally simple. If linear fittings are used, as with fluorescent lamps, it is likely that the light distribution will be different endwise from crosswise. A simple conversion factor makes allowance for the difference if the installation is viewed endwise and crosswise.

Again, a conversion term is available to allow for the fact that if the floor has a high reflection factor, it will add to the background luminance in the room and so reduce the degree of glare.

The whole procedure of computing the Glare Index for a complete installation has therefore been reduced to linear arithmetic and is consequently

* Tables of conversion factors relating to metric dimensions are given in the Appendix to IES Technical Report No. 10.

very simple. The time-consuming feature of the application of the empirical glare formula is incorporated in the precomputed tables of Initial Glare Index, so that the lighting designer has nothing more to do than a simple substitution in the tables with simple arithmetic conversion factors where necessary.

The Glare Index and Glare Sensation

The values of Glare Index determined from the IES tables are related to the glare constant from the empirical formula in the following way:

$$\text{Glare Index} = 10 \log_{10} (\text{glare constant}) \qquad (4.3)$$

If the values of the glare constant corresponding to each of the four basic criteria of glare (in the published experimental work by Hopkinson and his colleagues) are adjusted to equal 10, 40, 160 and 640 for the sensations of just perceptible, just acceptable, just uncomfortable, and just intolerable discomfort glare respectively (the original values taken directly from the experimental investigation were 8, 35, 150 and 600 as in Table 4.1 above); one obtains a series of glare constants from a starting point of 10, with a common ratio of 4 for each step in glare discomfort. The values of Glare Index on this basis therefore, by Eqn 4.3, correspond to 10, 16, 22, and 28 (10 log 10, 10 log 40, 19 log 160, 10 log 640) for the criteria just perceptible, just acceptable, just uncomfortable, and just intolerable respectively.

In further experiments (Collins 1962) in which observers were asked to make direct pair comparisons between complete installations, the results, when evaluated statistically, showed that one unit on the proposed scale of Glare Index corresponded to a just noticeable distinction between one installation and another. These studies also showed that three units on the proposed Glare Index scale represented a difference which would always be recognised and was therefore considered to be the difference which could be taken as a worthwhile improvement in glare amelioration.

Assessment of Limits of Acceptable Glare in Installations

The Glare Index limits to be used in the lighting code were determined by field studies. Teams of observers, whose ability to make subjective estimates of glare had been demonstrated under controlled conditions, visited a large number of installations and assessed the glare in terms of a scale which was directly related to the multiple criterion scale used in the laboratory experiments. Several series of appraisals were made, but in the final appraisals, from which the Glare Index limits for the IES Code were determined, the observing team were asked to specify whether the glare in an installation was acceptable or unacceptable for the particular purpose. In making this judgment, observers took into account the nature of the environment, the duration of time which people would spend in the room, the degree of freedom which they had to turn their gaze away from any glaring light source, and the degree of attention which the work demanded, which might either result in annoyance from the distraction caused by a glare source, or which might alternatively so command attention that the awareness of the environment might be reduced. These estimates of acceptability or unacceptability were related to the computed values of Glare Index obtained from the photometric and geometric constants of the installation. Limiting values of Glare Index were then allocated from a consideration of all these results.

The principle adopted in setting the limiting Glare Index values depended on broad categorisation. There was, in addition, the overriding decision that non-industrial environments would have stricter regulation of glare. Three broad categories were:

1. Environments where no glare at all is permissible; Glare Index limit 10.
2. Environments were glare must be kept to a minimum; Glare Index limit 13.
3. Environments where glare of different degree can be permitted depending upon the nature of the work, the likely sensitivity of people (children, elderly workers, sick people) and the time to be spent in the room, together with the degree of attention demanded by the work; Glare Index limits 16–28.

This method of assessing limiting values of Glare Index has a number of advantages. The whole system is selfcontained and is not related in any explicit way to a scale of sensations, although the sensation scale is implicit. Thus any criticism which might arise by the assignment of a degree of glare which in the laboratory would be designated 'just intolerable' to some particular industrial location is avoided by this method; if a Glare Index limit of 28 (corresponding to 'just intolerable' glare in the laboratory) is assigned to such a location, it does not mean that intolerable glare will, in fact, be experienced in the installation, but simply that a careful field appraisal coupled with past experience has indicated that such an installation with a Glare Index up to this limiting value of 28 will not be expected to give undue cause for complaint in practice.

There is also the advantage in the method in that if the values of Glare Index had been directly related to describe glare sensations, inevitably some experienced and sensitive individuals would detect differences between the indicated glare sensation level and their own actual sensation level. These discrepancies could well arise as a result of any of the simplifications introduced in the translation of the laboratory data into a practical method of routine evaluation.

It is of interest that in the United States of America proposals have been made to specify glare limits in terms of a probability function, that is, an index is given of the probable percentage of the general population which will experience glare below the borderline of comfort and discomfort (BCD). Thus an index of 85 means that 85% of the population will not experience glare beyond this borderline of comfort and discomfort. The disadvantage of this procedure as compared with the British system is again the explicit nature of the quoted value. Any group of people can evaluate an installation and demonstrate to their own satisfaction that, for example, a greater proportion of them experiences discomfort than would be indicated by the visual comfort index. This is always likely in any small specialised sample, for example, a group of doctors assessing the glare in a hospital ward. The discrepancy between the assessment and the prediction then becomes clear and obvious. With the British system, on the other hand, any such discrepancy only becomes significant if there are a significant number of complaints of glare discomfort in the installation. If this happens, and if the complaints are consistent in a number of similar environments, the matter can be rectified for future practice on the basis of this experience by altering the limiting value of Glare Index for that type of environment in the next edition of the IES Code. Such an arrangement, although it might appear to be concealing the situation, is by no means dishonest, any more than is a code which purports to lay down as absolute, standards which are related to highly variable individual reactions and which may well change with changes in standards of other aspects of the environment.

Alternative Methods of Determining Glare Index

The IES Glare Index system is designed for large installations of lighting fittings arranged in symmetrical arrays. Other methods are necessary where such an arrangement does not apply. Clearly in simple installations with only one or two light sources, the Glare Index can quite simply be obtained from the empirical formula. The calculation can, however, be aided by means of a suitable nomogram. Such a nomogram is shown in Fig. 4.3. In order to operate the nomogram, it is necessary to know the luminance of each individual source (in ft-L or $cd/m^2 \times 0{\cdot}292$) and the apparent area of each source in steradians.

The position of each source in a field of view must also be known in terms of its vertical and horizontal displacement from the line of sight, an appropriate position factor p being found from Table 4.4. (It will be seen that p is the ratio of the position index for the standard displacement—10° vertically above the line of sight—to the position index for the displacement of the source under consideration.) The luminance of the source and its apparent size can be measured or computed from the dimensions and from the polar curve of light distribution.

For the majority of lighting fittings which do not differ by more than 100 to 1 over the luminous surface, the source luminance can be found from the formula:

$$\text{Luminance} = \frac{\text{Intensity}}{\text{Projected luminous area in direction of view}}$$

Table 4.4

GLARE SOURCE POSITION FACTOR (p), WHERE $p = \left[\dfrac{\text{POSITION INDEX } (10°, 0°)}{\text{POSITION INDEX } (\theta°, \phi°)}\right]^{1.6}$

Horizontal Angle ($\phi = tan^{-1}\ L/R$)

Vertical Displacement (V/R)	0°	6°	11°	17°	22°	27°	31°	35°	39°	42°	45°	50°	54°	58°	61°	63°	68°	72°	Vertical Angle ($\theta = tan^{-1}(V/R)$)
1·9	—	—	—	—	—	—	—	—	—	0·02	0·02	0·02	0·02	0·02	0·02	0·02	0·02	0·02	62°
1·8	—	—	—	—	0·02	0·02	0·02	0·02	0·02	0·02	0·02	0·02	0·02	0·02	0·02	0·02	0·02	0·02	61°
1·6	0·03	0·03	0·03	0·03	0·03	0·03	0·03	0·03	0·03	0·03	0·03	0·03	0·03	0·03	0·03	0·03	0·03	0·03	58°
1·4	0·04	0·04	0·04	0·04	0·04	0·04	0·04	0·04	0·04	0·04	0·04	0·04	0·04	0·04	0·04	0·04	0·03	0·03	54°
1·2	0·05	0·05	0·06	0·06	0·06	0·06	0·06	0·06	0·06	0·06	0·06	0·05	0·05	0·05	0·05	0·04	0·04	0·04	50°
1·0	0·08	0·09	0·09	0·10	0·10	0·10	0·10	0·09	0·09	0·09	0·08	0·08	0·07	0·06	0·06	0·06	0·05	0·05	45°
0·9	0·11	0·11	0·12	0·13	0·13	0·12	0·12	0·12	0·12	0·11	0·10	0·09	0·08	0·07	0·07	0·06	0·06	0·05	42°
0·8	0·14	0·15	0·16	0·16	0·16	0·16	0·15	0·15	0·14	0·13	0·12	0·11	0·09	0·08	0·08	0·07	0·06	0·06	39°
0·7	0·19	0·20	0·22	0·21	0·21	0·21	0·20	0·18	0·17	0·16	0·14	0·12	0·11	0·10	0·09	0·08	0·07	0·07	35°
0·6	0·25	0·27	0·30	0·29	0·28	0·26	0·24	0·22	0·21	0·19	0·18	0·15	0·13	0·11	0·10	0·10	0·09	0·08	31°
0·5	0·35	0·37	0·39	0·38	0·36	0·34	0·31	0·28	0·25	0·23	0·21	0·18	0·15	0·14	0·12	0·11	0·10	0·09	27°
0·4	0·48	0·53	0·53	0·51	0·49	0·44	0·39	0·35	0·31	0·28	0·25	0·21	0·18	0·16	0·14	0·13	0·11	0·10	22°
0·3	0·67	0·73	0·73	0·69	0·64	0·57	0·49	0·44	0·38	0·34	0·31	0·25	0·21	0·19	0·16	0·15	0·13	0·12	17°
0·2	0·95	1·02	0·98	0·88	0·80	0·72	0·63	0·57	0·49	0·42	0·37	0·30	0·25	0·22	0·19	0·17	0·15	0·14	11°
0·1	1·30	1·36	1·24	1·12	1·01	0·88	0·79	0·68	0·62	0·53	0·46	0·37	0·31	0·26	0·23	0·20	0·17	0·16	6°
0	1·87	1·73	1·56	1·36	1·20	1·06	0·93	0·80	0·72	0·64	0·57	0·46	0·38	0·33	0·28	0·25	0·20	0·19	0°
Lateral Displacement (L/R)	0	0·1	0·2	0·3	0·4	0·5	0·6	0·7	0·8	0·9	1·0	1·2	1·4	1·6	1·8	2·0	2·5	3·0	

V = Vertical distance from horizontal line of vision
L = Lateral distance from horizontal line of vision
R = Horizontal distance from eye

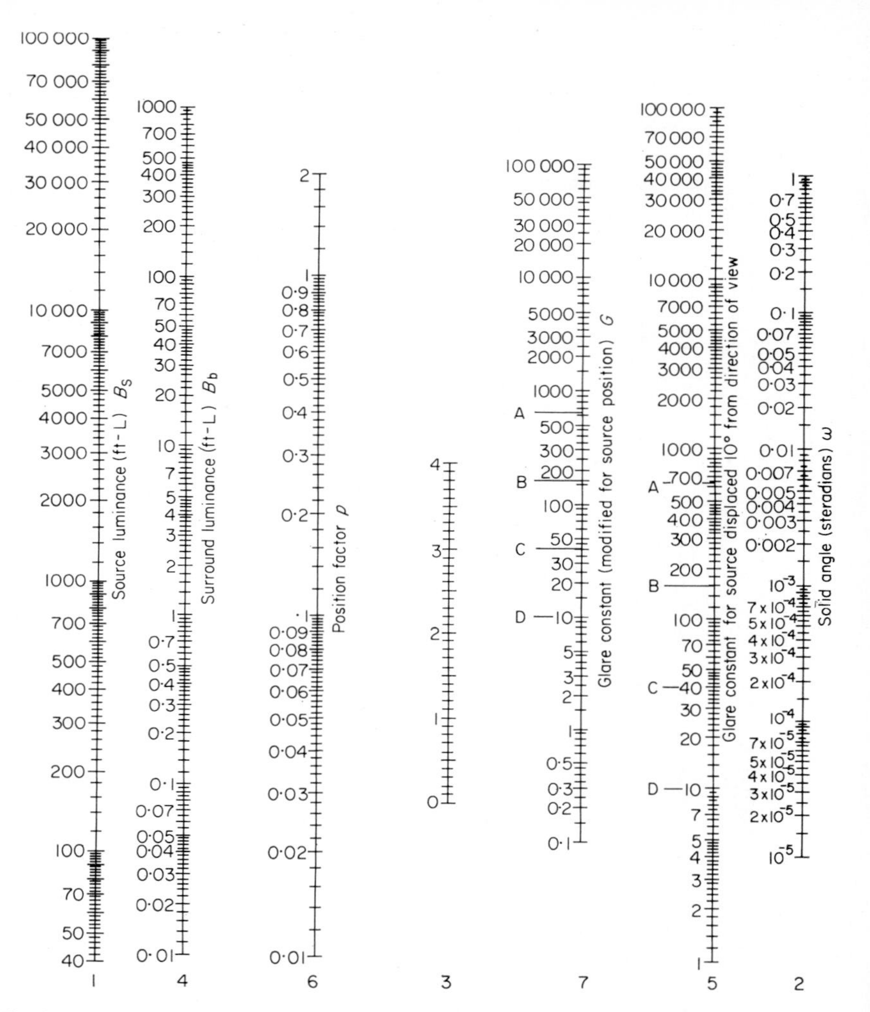

Fig. 4.3 *Nomogram for the calculation of glare constant for nonsymmetrical layouts from the BRS glare formula* $G = \dfrac{B_s^{1.6}\,\omega^{0.8}}{B_b}\,p$

To operate the nomogram a straight line is first drawn from the value of source luminance (ft–L, or cd/m² × 0·292) on Scale 1 of the nomogram to the value of source solid angle (steradians) on Scale 2, and the intercept on Scale 3 is found. Through this intercept on Scale 3, a line is drawn through the value of the surround luminance (ft–L or cd/m² × 0·292) on Scale 4 to meet Scale 5. The value on Scale 5 is the glare constant for a source displaced 10° vertically above the direction of view. A line can then be drawn between this value of glare constant on Scale 5 to the value on Scale 6 of the Position Factor as determined from Table 4.4. The glare constant modified for source position is then given by the intercept on this line on Scale 7.

The intensity in the direction of the observer is found from the polar curve of light distribution supplied by the manufacturer.

Alternatively, the luminance of the fitting can be measured directly with a luminance meter; if the fitting is of nonuniform luminance, the arithmetic average within the boundaries of the fitting can be taken as the luminance for substitution in the nomogram.

The solid angle subtended by the light source can be taken from the linear dimensions of the fitting and its distance from the observer using the formula:

$$\text{Solid angle} = A \cos\theta \cos\phi/d^2$$

where A is the luminous area of the fitting projected on a vertical or horizontal plane parallel to an axis of the fitting not parallel to the line of sight, θ and ϕ respectively the angles between the normal to the plane of projection and the direction of the source from the observer in the vertical and horizontal planes and d the distance of the source from the observer expressed in the same units as the area A (see Fig. 4.4). A visual gauge designed by Petherbridge and Longmore (1954) and illustrated in Fig. 4.5 enables this quantity to be measured directly.

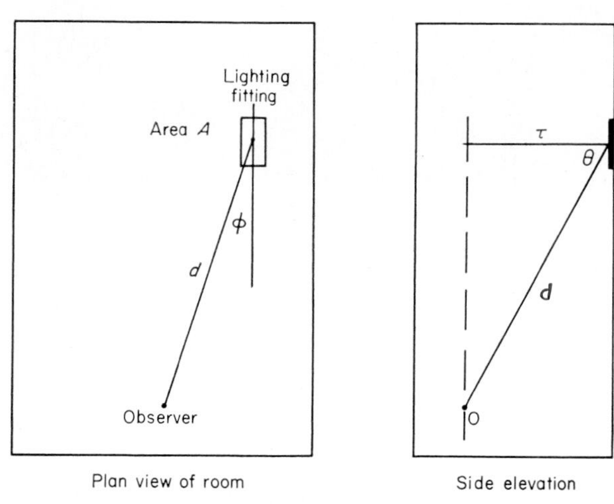

Fig. 4.4 *Solid angle subtended by fitting* = $A \cos\theta \cos\phi/d^2$

It is then necessary to obtain the average surround luminance in the room. In an existing installation, this can be taken as the numerical equivalent, expressed in foot-lamberts, of the illumination on the eye of the observer from the whole of the room, less the illumination from the fittings themselves. Alternatively the indirect component of illumination in the installation can be computed in lm/ft^2 (or lux/10·76), and the average surround luminance can be taken as the numerical equivalent expressed in foot-lamberts.

The glare constant for each particular lighting fitting in the installation is computed as indicated in Fig. 4.3, and the values of glare constant for every fitting in the field of view so obtained are then added arithmetically.

The value of the Glare Index is then given by the formula:

$$\text{Glare Index} = 10 \log_{10} (\text{arithmetic sum of all glare constants})$$

Fig. 4.5 *The BRS solid-angle gauge*

This procedure causes little difficulty in a small installation of only a few sources. In practice, a number of simplifications can be introduced. If there are a number of sources, but they are not arranged symmetrically to permit the use of the IES Glare Index tables, it is often possible to calculate the glare constant for a few selected fittings with the aid of the nomogram, and thence to interpolate or extrapolate from these values to arrive at a sufficiently close estimate of the values of glare constant for the other remaining fittings. In the same way, a closely arranged array of lighting fittings, or a multilight fitting of the decorative type, can often be equated to a single fitting located at the centroid of the system, with a glare constant of, say, ten times the glare constant for each of the ten component parts. Common-sense and experience will quickly indicate how these various short cuts can be made.

Validity of the Glare Index Values

It must be stressed that the values of IES Glare Index, or the values of Glare Index obtained from individual computations or with the use of the nomogram, are only valid within the range of experimental conditions which led to the empirical formula upon which the system is based. Situations which involve luminous areas of high luminance and large solid angle cannot be used together. For this reason the operative range of the nomogram has been limited and it is unwise to extrapolate any of the scales for combinations of sourcc luminance and source solid angle which fall outside the limits of the scales. Equally, the IES Glare Index tables cannot be extrapolated, although it is legitimate to interpolate linearly within them for conditions which are not specifically tabulated.

Even so, it is possible to find situations beyond the experimental validity of the Glare Index system within the confines of the nomogram or of the IES tables. Without going into great detail, it is only possible to give a warning that since the system is based upon an empirical formula which is itself a simplification of a highly complex situation revealed by empirical experiments (and not a validated theory of vision), the greatest care should be taken not to stretch the operation of the system without full realisation of the errors which might result. It is easy, of course, to condemn the whole system by reason of its breakdown at extremes, but opinion in this country is satisfied that the correct balance has been achieved with the Glare Index system between one which is simple to operate and one which is universally applicable.

Additional Factors in Glare Control

It will be evident that glare control with the IES system depends essentially upon limiting the luminance of the light sources as seen by the observer, and/or limiting their size, and providing as high a background luminance as possible to buffer the glare effects of the sources. There are other ways of ameliorating glare which are, however, of minor importance compared with these major factors. One method of mitigating glare is that known as 'contrast grading'. It has been shown experimentally (Petherbridge and Hopkinson 1950, Hopkinson 1951) that a glaring source which is not sharply delineated against the background causes less glare than a sharp edge. For example, an opal sphere enclosing a lamp and having uniform luminance is more glaring than a fitting of the same total area in which the centre is of higher brightness and there is a gentle gradation towards the edge. In practice, suitable contrast grading can result from a design of fitting in which semi-diffusers rather than complete diffusers are used. Alternatively, the light distribution from the fitting can be so arranged to provide an immediate surround on the ceiling or adjacent wall such that the immediate background to the lighting fitting as seen by the observer has a higher luminance than the rest of the surroundings, giving a gradation of luminance from the brightest central part of the fitting to the general surroundings. Such careful contrast grading design is always worth considering especially in a marginal installation where straightforward methods of lowering the Glare Index are not possible. At the moment the IES Glare Index system does not take contrast grading into account. Consequently the use of contrast grading in design does not help the situation as far as meeting statutory limits based on the IES Code is concerned. The technique is rather one for the discerning designer to use to demonstrate refinements of design to achieve an effect which has superiority over conventional methods.

Indirect Glare from Veiling Reflections

Direct glare from lighting fittings is the greatest source of discomfort in artificial lighting, but sometimes the measures taken to reduce direct glare have an unfortunate effect upon reflected indirect glare. Indirect glare is caused by reflections of light sources in polished table tops, machinery, or other glossy surfaces, or by the sheen on glossy or semiglossy paper or other working surfaces. These various reflections can cause discomfort or interference with vision in a number of ways. The specular images of light fittings in polished surfaces may cause direct glare discomfort themselves, or they may cause distraction. Again, the sheen effect can materially reduce the contrast of the visual task and so lower visual performance. Different effects can be present together.

If the visual task is brighter than the surroundings, the attention is drawn to, and held on, the task with the minimum of conscious effort. On the other hand, if the surroundings are brighter than the task itself, or particularly if there are any areas visible in the immediate surround to the task which have a luminance greater than that of the task itself, these will have a distracting effect (see Chapter 8). When this latter condition arises, the worker may be under continual strain in trying to maintain attention on his work in the face of the distraction, and may well become prone to feelings of tiredness or fatigue.

Distraction caused by bright images in the field of view near the visual task is one of the main effects of reflected glare, particularly if the images are fairly sharp but not in the same plane as the task. This distraction arises not least because the virtual image of the light source is not in the same plane as that of the work, and so the image as seen by the two eyes is doubled when the observer is looking at his work, but becomes fused into a single image when he looks slightly away into the polished table top and allows his convergence to relax to the distance of the virtual image. This continual process of convergence and relaxation around a virtual image can be distressing. Petherbridge and Hopkinson (1955) showed that, although the luminance of such reflected images might not be sufficiently high to cause actual glare discomfort, the distraction could be a serious problem. The remedy is to avoid glossy surfaces in the region of the task or the working plane, and to avoid placing the work in relation to light sources so that such specular reflections arise.

The reduction of contrast of the task due to veiling reflectances can also be serious. Pencil lines, printed characters and so on lose contrast if a light source or bright area is reflected preferentially in their vicinity. The dials of meters on a panel, particularly if they are seen behind a glass cover, are obviously vulnerable. Aspects of veiling reflectance have been studied in detail by Finch, Chorlton and Davidson (1959) with particular reference to school and office tasks. They evaluated the loss in visibility due to veiling reflections on the work. Blackwell (1963A, B) has carried out investigations with polarising screens on lighting fittings as a means of ameliorating veiling reflections.

The Australian Standards Association (1957), advised by the Australian Illuminating Engineering Society, has made recommendations for the distribution of lighting in order to avoid veiling reflections. The Australian lighting code recommends that where veiling reflections are likely to affect visibility, the illumination, expressed in lm/ft^2, should be numerically not less than twelve times the luminance of the light fittings expressed in $candelas/in^2$ (or in lux, not less than $\frac{1}{13}$th of the luminance in cd/m^2). This recommendation ensures that the direct illumination component will be sufficiently great in relation to the 'sheen' component, to ensure adequate contrast between the visual task and its background. The recommendation is for particular application where the task consists of tracing drawings on to shiny linen tracing cloth, but the recommendations are of general application.

Griffith (1964) has demonstrated the advantages of side lighting for the reduction of veiling glare and contrast degeneration, as compared with conventional top lighting. It is often possible by a suitable combination of top lighting together with a direct component from the side so to swamp veiling reflections to achieve a result sufficiently satisfactory for all but the most critical tasks. In his proposals for a lighting code based on luminance, Hopkinson (1965) proposes that the selective component of lighting on the work should have directional properties and be so combined with the general building lighting that not only are satisfactory luminance ratios achieved, but unwanted veiling reflections are reduced or eliminated.

References

Australian Standards Association (1957). Australian Standard Code for the Artificial Lighting of Buildings. AS–CA30–1965. Standards Association of Australia, Sydney.

Bartlett, F. C. and K. G. Pollock (1935): Industrial Health Research Board Report No. 65. HMSO, London.

Blackwell, H. R. (1963A): 'A General Quantitative Method for Evaluating the Visual Significance of Reflected Glare.' *Illum. Engng.* (New York), **58**, 161–216.

Blackwell, H. R. (1963B): 'A Recommended Field Test Method for Evaluating Overall Visual Efficiency of Lighting Installations.' *Illum. Engng.* (New York), **58**, 642–647.

Christie, A. W. and A. J. Fisher (1966): 'The Effect of Glare from Street Lighting Lanterns on the Vision of Drivers of Different Ages.' *Trans. Illum. Engng. Soc.* (*London*), **31**, 93–108.

Collins, W. M. (1962): 'The Determination of the Minimum Identifiable Glare Sensation Interval.' *Trans. Illum. Engng. Soc.* (*London*), **27**, 27–34.

Crawford, B. H. and W. S. Stiles (1935). 'A Brightness Difference Threshold Meter for the Evaluation of Glare from Light Sources.' *J. Sci. Instrum.* **12**, 177–185.

Crawford, B. H. and W. S. Stiles (1937): 'The Effect of a Glaring Light Source on Extrafoveal Vision.' *Proc. Roy. Soc. B.*, **122**, 255–280.

Einhorn, H. (1961): 'Predetermination of Direct Discomfort Glare.' *Trans. Illum. Engng. Soc.* (*London*), **26**, 154–164.

Finch, D. M., J. M. Chorlton and H. F. Davidson (1959): 'The Effect of Specular Reflection on Visibility.' *Illum. Engng.* (*New York*), **54**, 477–488.

Fugate, J. and G. A. Fry (1956): 'Relation of Changes in Pupil Size to Visual Discomfort.' *Illum. Engng.* (*New York*), **51**, 537–549.

Griffith, W. (1964). 'Analysis of Reflected Glare and Visual Effect from Windows.' *Illum. Engng.* (*New York*), **59**, 184–188.
Guth, S. K. (1963): 'A Method for the Evaluation of Discomfort Glare.' *Illum. Engng.* (*New York*), **58**, 351–364.
Holladay, L. L. (1926): 'The Fundamentals of Glare and Visibility.' *J. Opt. Soc. Amer.*, **12** 271–319.
Hopkinson, R. G. (1940): 'Discomfort Glare in Lighted Streets.' *Trans. Illum. Engng. Soc.* (*London*), **5**, 1-30.
Hopkinson, R. G. (1951): 'The Brightness of the Environment and its Influence on Visual Comfort and Efficiency.' Proc. Building Research Congress, Div. 3, Part III, 133–138.
Hopkinson, R. G. (1955): 'Subjective Judgments—Some Experiments Employing Experienced and Inexperienced Observers.' *Brit. J. Psychol.*, **46**, (Pt. 4), 262–272.
Hopkinson, R. G. (1956): 'Glare Discomfort and Pupil Diameter.' *J. Opt. Soc. Amer.*, **46**, 694–656.
Hopkinson, R. G. (1957): 'Evaluation of Glare.' *Illum. Engng.* (*New York*), **52**, 305–316.
Hopkinson, R. G. (1965): 'A Proposed Luminance Basis for a Lighting Code.' *Trans. Illum. Engng. Soc.* (*London*), **30**, 63–88.
Hopkinson, R. G. and P. Petherbridge (1954): 'Two Supplementary Studies on Glare.' *Trans. Illum. Engng. Soc.* (*London*), **19**, 220–224.
Illuminating Engineering Society, London (1968): The IES Code: 'Recommendations for Good Interior Lighting.'
Illuminating Engineering Society, London (1967): Technical Report No. 10: 'Evaluation of Discomfort Glare: The IES Glare Index System for Artificial Lighting Installations.'
Ivanoff, A. (1947): 'The Inhibitory Component of Glare.' *Rev. d'Optique*, **26**, 479–488.
Luckiesh, M. and S. K. Guth (1949): 'Brightness in Visual Field at Borderline Between Comfort and Discomfort (BCD).' *Illum. Engng.* (*New York*), **44**, 650–670.
Luckiesh, M. and L. L. Holladay (1925): 'Glare and Visibility.' *Trans. IES* (*New York*), **20**, 221.
Petherbridge, P. and R. G. Hopkinson (1950): 'Discomfort Glare and the Lighting of Buildings'. *Trans. Illum. Engng. Soc.* (*London*), **15**, 39–79.
Petherbridge, P. and R. G. Hopkinson (1955): 'A Preliminary Study of Reflected Glare.' *Trans. Illum. Engng. Soc.* (*London*), **20**, 255–257.
Petherbridge, P. and J. Longmore (1954): 'Solid Angles Applied to Visual Comfort Problems.' *Light and Lighting*, **47**, 173–177.
Schouten, J. F. (1937): 'Visual Measurements of the Adaptation and Mutual Influence of the Retinal Elements.' Thesis, Utrecht.
Schouten, J. F. (1938): 'The Rotating Pendulum and the State of Adaptation of the Eye.' *Nature*, **142**, 615.
Stiles, W. S. (1928–9): 'The Effect of Glare on the Brightness Difference Threshold.' *Proc. Roy. Soc. B.*, **104**, 322–351.
Stiles, W. S. (1929): 'The Scattering Theory of the Effect of Glare on the Brightness Difference Threshold'. *Proc. Roy. Soc. B.*, **105**, 131–146.
Stiles, W. S. (1930): 'A Brightness Threshold Meter (or Glaremeter).' *Illum. Engr.* (*London*), **23**, 279–280.
Stiles, W. S. (1931): 'Evaluation of Glare in Street Lighting Installations, Parts I and II.' *Illum. Engr.* (*London*), **24**, 162–166, 187–189.
Stiles, W. S. (1935): 'Comparison of the Revealing Powers of White and Coloured Headlight Beams in Fog.' *Illum. Engr.* (*London*), **28**, 125–132.
Stone, P. T. and S. D. P. Groves (1968): 'Discomfort Glare and Visual Performance.' *Trans. Illum. Engng. Soc.* (*London*), **33**, 9–15.

5

Intermittency and Flicker

The first supplies of electricity for lighting used direct current, but when systems of transforming alternating current to higher voltages for more economical transmission were developed, the problem of flicker in lighting first came into prominence.

In any lamp, the instantaneous power being consumed is proportional to I^2R, where I is the instantaneous current, and R the resistance (or in-phase impedance), so that when fed with alternating current in one whole cycle there will be two positive waves of power (Fig. 5.1). The frequency of the power cycle is therefore double that of the supply frequency, and so in Europe, where the standard supply frequency is 50 Hz, the lamp power fluctuates at 100 Hz. In North America the standard frequency is 60 Hz, the lamp power frequency then being 120 Hz. (1 hertz (Hz) = 1 cycle per second.)

The extent to which the fluctuation of light output follows the fluctuation of power in a lamp will depend on the inertia of the light-producing system. In a filament lamp the thermal capacity of the filament has a major effect on the extent of the fluctuations. A low-current lamp with a thin filament has a lower thermal capacity than a heavy-current lamp with a thick filament, and therefore the fluctuations of light output follow to a greater extent the fluctuations of power. Lamp flicker is therefore potentially a more noticeable phenomenon in Great Britain, which uses a higher mains voltage (240 V) and a lower frequency (50 Hz) than in North America which uses a lower mains voltage (110 V) and a higher frequency (60 Hz), for two reasons, (a) because an incandescent filament lamp of the same wattage will have a thinner filament and therefore a lower thermal capacity, and (b) the frequency is lower.

In a gaseous discharge lamp (and this includes fluorescent tubes) the light is not produced by incandescence but by an ionisation process which is very rapid in comparison with the power fluctuations, and therefore the light output follows the power fluctuations very closely if the discharge itself is the major source of light emission. With the fluorescent tube, however, while the major part of the *energy* emitted by the lamp is in the ultraviolet region of the spectrum, the *visible light* from the lamp comes from the luminescent material (or 'phosphor coating') on the walls of the tube, and the inter-

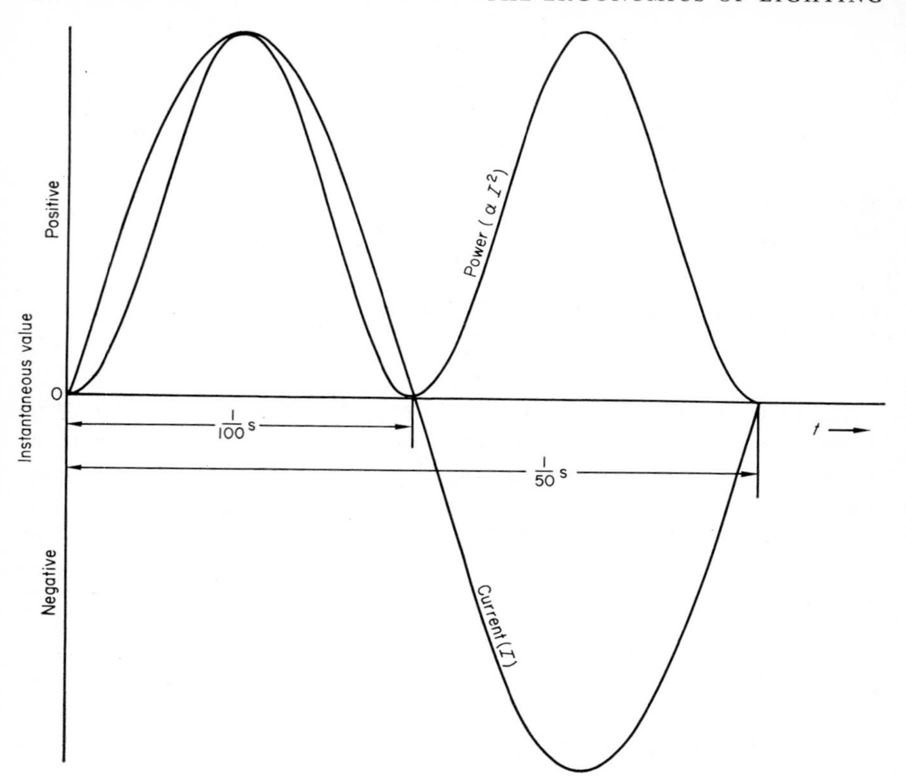

Fig. 5.1 *Relation between current and power frequencies*

mittency of this light depends upon the 'afterglow' of this coating. Many of the luminescent materials used in fluorescent lamps have different characteristics in respect of the rapidity with which the emitted light follows the absorbed ultraviolet, and while some phosphors have a short time constant so that the emitted light follows very closely the fluctuations in energy of the discharge, others have a relatively long afterglow which operates much in the same way as the thermal inertia of the incandescent filament. Consequently the amplitude of the fluctuations of light output in such fluorescent lamps is very much less than that of the discharge itself.

The visual effect of a light which is fluctuating may be perceived in one or both of two ways. The light itself may not be perceived directly as flickering, but it may be noticed that objects which are moving appear to move in jerks, while rotating machinery may have a more or less blurred pattern superimposed which may be stationary or moving slowly in the same or even the opposite direction of rotation to the true motion. This is the well-known 'stroboscopic' effect.

On the other hand, if the frequency of the fluctuation is low, the actual fluctuations may be perceived by the eye either in the light source itself or in the illuminated visual field. Both these effects, the stroboscopic effect, and the direct flicker effect, have become of renewed interest since the introduction of fluorescent lighting. Normally they are not serious problems, but they can sometimes give trouble. Techniques have been devised, following basic experiments on human vision, to minimise or eliminate the effects in good lighting design.

The subject of flicker perception has long been of interest because the critical fusion frequency, that is, the frequency at which perception of flicker just disappears, is known to offer a method of determining the time constants of certain parts of the visual mechanism. The CFF, as it is called, is therefore one of the major parameters of visual research. Long before discharge lamps came into general use, Porter (1902) determined a relation between the CFF and the luminance (brightness) of the source. Lythgoe and Tansley (1929) related the adaptation characteristics of the eye to the CFF. It was not, however, until the advent of techniques of electroretinography, micro-electrode work on the neural fibres of the retina, and electroencephalography, that it was possible to determine to what extent the fusion of a train of light impulses arriving on the retina occurred in the retina itself, in the nerve systems, or in the higher centres of the brain. Ireland (1950) demonstrated by binocular interaction effects that fusion must occur mainly in the higher centres. More recently it has been shown that quite high frequencies of signals can be transmitted up the optic nerve to these higher centres, and one school of thought insists that the effect of these high-frequency signals arriving from the peripheral organ must give rise to extra loading of the central nervous system, and hence that this must account for some of the complaints of an earlier onset of feelings of tiredness with lighting from an intermittent source such as fluorescent lighting.

It is a matter of common clinical observation and, in fact, a part of current clinical technique, that intermittent stimulation by light with a frequency of around 10 Hz, corresponding to the frequency of the characteristic wave found in the electroencephalogram of a resting subject (the 'alpha-rhythm') can produce nervous stimulation, and if the subject has any epileptic tendencies, can provoke a seizure. There exists also a literature of seizures or spasms provoked in such subjects by television screens when badly adjusted, or even by fluorescent lamps when the lamp is malfunctioning, for example, when end-flicker is obtrusive. Other records exist of seizures provoked in people cycling or driving down avenues of trees, due presumably to a critical rhythm of light fluctuation having thus been produced. None of this should be construed as evidence that flicker from fluorescent lighting is a danger or is even mildly harmful to people with normal vision and a properly functioning nervous system. On the other hand, it serves no useful purpose to deny that flicker exists in fluorescent lighting or that it can, on occasions, be a nuisance to the normal population, and a disturbance and a distress to people with quite minor abnormalities. People who suffer from symptoms arising from fluctuating light sources, whether from fluorescent lighting or watching the television or cinema screen, should recognise their disability and avoid placing themselves in situations which cause them trouble.

Since the fluctuation of light from fluorescent lamps has its major component at 100 Hz, it was originally claimed that since this frequency was well above the CFF as determined by the classical investigators, there was no likelihood of fluorescent lighting giving rise to perceptible flicker in a lighting installation. However, during the initial period of the spread of this type of lighting in the decade after the lamp was first commercially available on a large scale, a small but persistent amount of complaint led to a detailed investigation of the problem of flicker at the Building Research Station (Collins and Hopkinson 1954, 1957; Collins 1956A, B).

Before this investigation was undertaken, much of the work on flicker perception had been carried out with small fields of view, and at moderately low luminance levels. It was one of the aims of the new investigation to extend this earlier work to the stimulation of the whole visual field exactly as in a normal lighting installation which used lamps of fluctuating light output such as fluorescent tubes. A further extension of earlier work was the use of a multiple criterion technique of making subjective judgments of the discomfort caused by flicker in addition to the CFF criterion of 'just imperceptible'.

The immediate objects of the investigation were to determine:

1. The variability of sensitivity to flicker of different people under large-field or full-field conditions.
2. The effect of frequency of flicker below the CFF value on the degree of discomfort caused.
3. The effect of increasing the apparent area of the fluctuating luminous field up to the maximum possible.
4. The effect of the luminance of the visual field, particularly for large visual fields.
5. The effect of waveform of the fluctuating source.

The studies of these factors were supplemented by measurements of the waveform of light output from various types of light sources and also from fluorescent lamps using various control circuits and arrangements of lamps. Frequency was always the variable parameter in the experiments (except for the subharmonic studies reported later), and in all cases the subject was first asked to report the point at which flicker first appeared as the frequency was lowered from a high value. After this, the frequency was lowered successively until the subject reported:

Just obvious flicker;
Just uncomfortable flicker, and finally:
Just intolerable flicker.

In several of the studies, the subjects were asked immediately afterwards to repeat the assessment of the criteria in the reverse order with the frequency increasing, in order to discover the magnitude of the adaptation effects. The results of these latter observations were not used for the determination of the main relationships, however, since they would not be applicable to normal lighting conditions.

The presentation of a visual field of which the brightnesses of all parts were fluctuating at the same frequency, and of which the frequency of fluctuation could be controlled, was achieved in two different ways in the course of the investigation. The first series of results were obtained with the apparatus shown in Fig. 5.2. The subject looked into a box and viewed a white screen uniformly illuminated and occupying a large area of the field of view. Towards the top of the field of view, a small slit was cut, and this slit was illuminated from behind to the luminance of a fluorescent lamp, to represent such a lamp in the field of view at a distance of 15 ft. The whole of this visual field was caused to fluctuate in luminance by a sectored disc mounted in front of the eyes of the subject, and the subject could control the speed of the motor himself in making the observations, the speed being measured stroboscopically by the experimenter.

This apparatus gave valuable results concerning the variability of sensitivity from subject to subject and of each subject from day to day over a long period; results were also obtained for bright fields of view of 50 ft-L (172 cd/m^2) subtending angles of 10° × 10° up to 30° × 60°. Other results (which were unexpected) were obtained which showed that there were effects of phase differences between the stimulation of the two eyes when viewing

Fig. 5.2 *Small-field flicker test apparatus*

binocularly, so that an experimental arrangement to illuminate the whole of the visual field was designed to avoid such phase differences.

The source of variable-frequency fluctuating light used to illuminate a small room consisted of two slotted drums, one of which was stationary and the other rotating, one inside the other, and a 1 kW Class B projector lamp running on DC, which was mounted inside the two cylinders. This fitting was mounted at the centre of a room of about 10 ft × 8 ft floor area (3 m × 2·5 m), and a large cylinder of diffusing polythene film surrounded the whole fitting and helped to ensure that the illumination of the walls of the room was uniform. The frequency of fluctuation of the light was controlled by the speed of the motor rotating the inner drum, and the speed of this rotation was measured by means of a tachometer generator connected to a rectifier voltmeter. A steady light source suspended below the fluctuating source was used to vary the modulation of the light waveform from 100% down to 10% by providing steady illumination of the room.

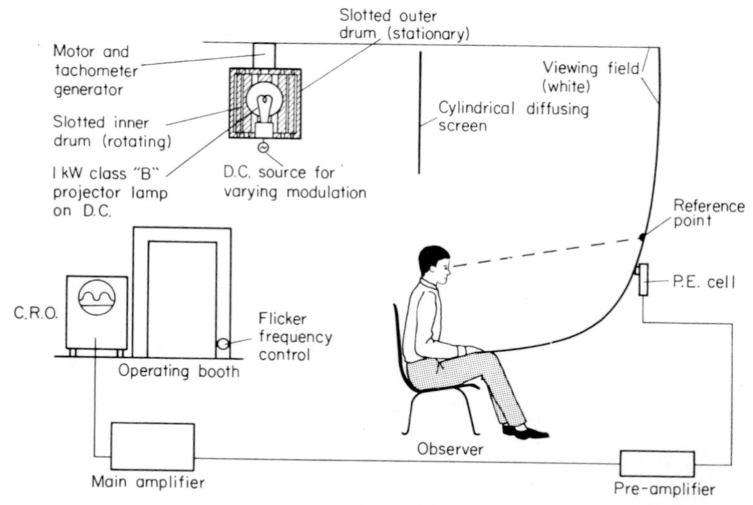

Fig. 5.3 *Experimental arrangement for full-field flicker observations*

The experimental arrangement is shown in Fig. 5.3. The subject observed a screen covering the end wall of the room which was arranged so that the whole visual field was at approximately the same luminance. The lighting fitting was not visible to the subject when making the observation, but two marks on the end wall at eye level and five degrees to his left and to his right gave him reference (but not fixation) points between which he was asked to allow his gaze to wander naturally. The monitoring photoelectric cell connected to the oscillograph was placed behind a translucent screen in the end wall.

The first question to be answered by the investigation was that of the difference between subjects in their perception of flicker or degree of dis-

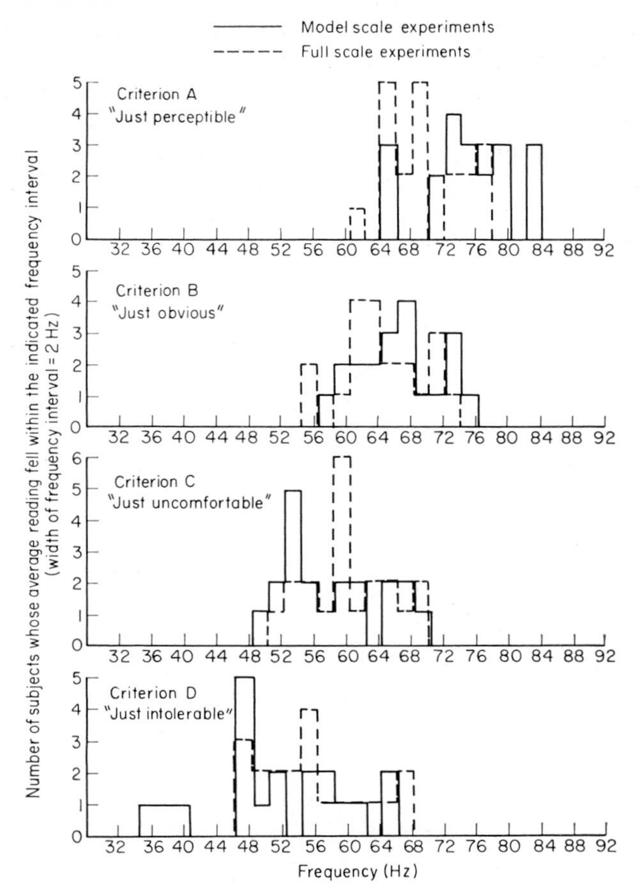

Fig. 5.4 *Distribution of average values of flicker frequency for 20 subjects, for four criteria of flicker sensation*

comfort felt. The answer to this will be seen in Fig. 5.4, which shows that there is a fairly wide spread of results from subject to subject for each of the four criteria, and that these spreads overlap, so that flicker of, say, 66 Hz, while only just perceptible to some subjects, could cause just intolerable discomfort to others. This result indicates the necessity of treating complaints of flicker on an individual basis; there is no justification for a foreman or manager in a factory to insist that because flicker is not apparent to him, a worker who complains of flicker must be irresponsibly inventing complaints. The converse is also true, of course; because one man is troubled by flicker, it does not necessarily mean that the whole of a workshop or office is being subjected to annoyance.

The results in Fig. 5.4 are based on the mean values of frequency obtained for the twenty subjects over at least six different determinations. To study how subjects varied in their sensitivity over a period of time, some of them repeated the control experiment at intervals throughout the period of the investigation. The way in which the frequencies chosen for each of the different criteria varied over a period totalling seven months is shown for one of the observers in Fig. 5.5. It will be seen that on any one day the subject would be expected to choose a frequency within ±7 or 8% of his mean value for that sensation. It will also be seen that the variations from day to day appear to be those of general sensitivity, i.e. the curves for the four different criteria are roughly parallel.

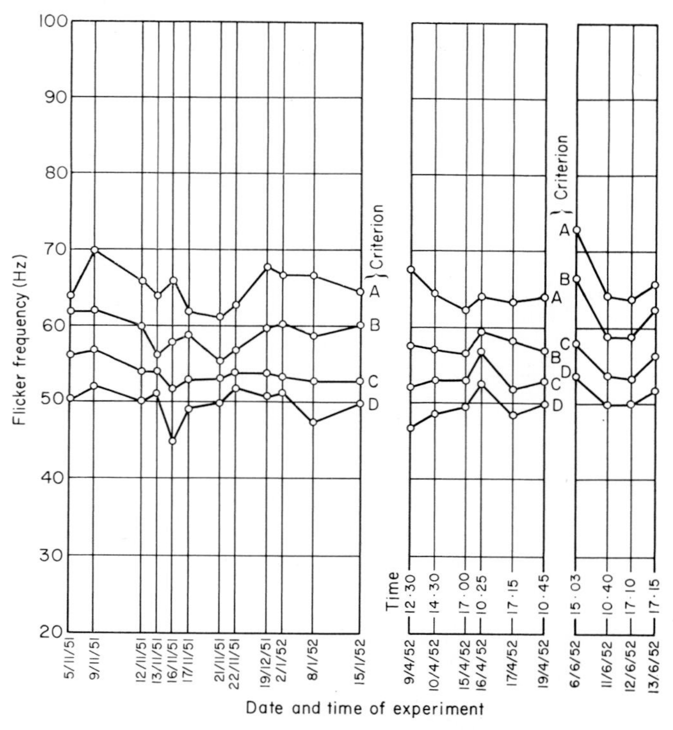

Fig. 5.5 *Variation of one subject on control experiment*

The total range of variation of frequency for the observers with the largest, smallest and the average range for each criterion are given in Table 5.1. This table also shows, apart from the actual average values of frequency and the standard deviation, the separation between each of the criteria. It is interesting to see that the four appear to be quite evenly spaced with a separation of about 0·06 log units for the model experiment and 0·03 for the full-field experiment. This means that, in the case of a complete lighting installation, a difference of 4 to 5 Hz will mean a change in appreciation of flicker from just obvious flicker to just uncomfortable flicker.

There were two major differences between the conditions of the model-scale and the full-scale experiment: the first was that the field luminance in the

Table 5.1

RESULTS OF OBSERVATIONS OF FLICKER SENSATION MADE BY 20 SUBJECTS

	Criterion	*Mean frequency for each criterion*			*Range of variation of subjects*			*Separation between each criterion and that following (log units)*		
		Hertz	*Std. Dev.*	*Coefficient of variation*	*Av. range*	*Smallest range*	*Largest range*	*Av. value*	*Smallest value*	*Largest value*
				Per cent.	*Per cent.*	*Per cent.*	*Per cent.*			
Model-scale experiment (50 ft–L) (170 cd/m²)	A Just perceptible	75·1	5·6	7·4	19·4	8	42	0·054	0·013	0·10
	B Just obvious	66·4	8·5	12·8	15	5	25	0·057	0·034	0·11
	C Just uncomfortable	58·2	6·1	10·5	15	7	25	0·064	0·028	0·165
	D Just intolerable	50·3	8·5	16·9	24	6	50			
All-round visual field experiment (5 ft–L) (17 cd/m²)	A Just perceptible	70·1	4·6	7	12·1	2·6	59·6	0·033	0·012	0·079
	B Just obvious	64·8	5·2	8	8·7	2·8	23·0	0·028	0·011	0·055
	C Just uncomfortable	60·9	5·7	9·3	9·4	4·5	18·2	0·037	0·013	0·068
	D Just intolerable	55·9	6·5	11·7	9·2	3·1	20·5			

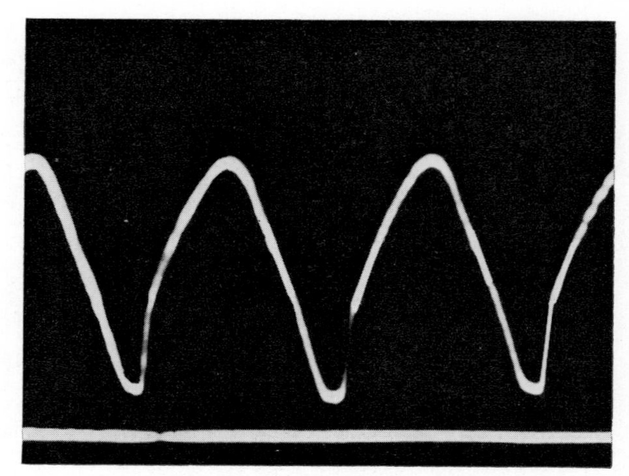

Fig. 5.6 *Light output waveform from 'cool' colour fluorescent lamp*

latter was only one-tenth of that in the former, which factor would normally lead one to expect, on the basis of the Ferry–Porter law, a considerable reduction in CFF. The second difference, however, that of increase of the field size by about 25 times, appears to compensate to a large extent for the drop in luminance and a drop of only 5 Hz in the 'just perceptible' results on average.

In addition to the luminance of the visual field and its size, and the frequency of the fluctuations, the waveforms of the latter have an important effect on the perception of flicker. With the square or rectangular or at least straight-sided waveforms usually obtained from mechanical laboratory apparatus, it is possible to differentiate between the dark/light time ratio of the fluctuations and their modulation (that is, the relation between the height

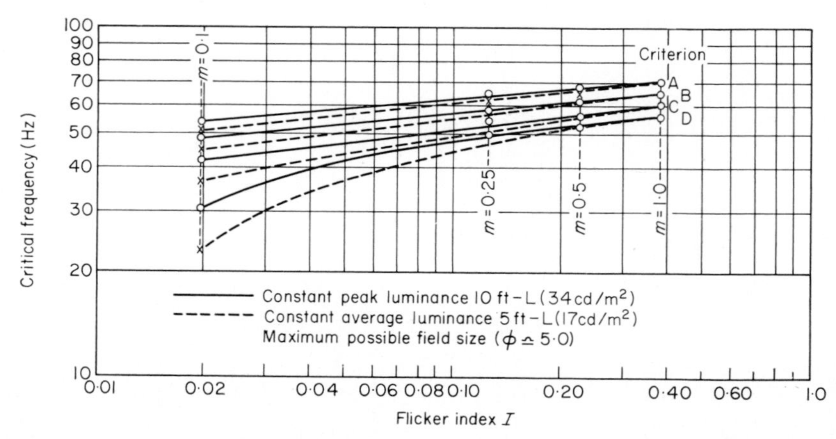

Fig. 5.7 *Relation between flicker index and critical frequency for four criteria of flicker sensation*

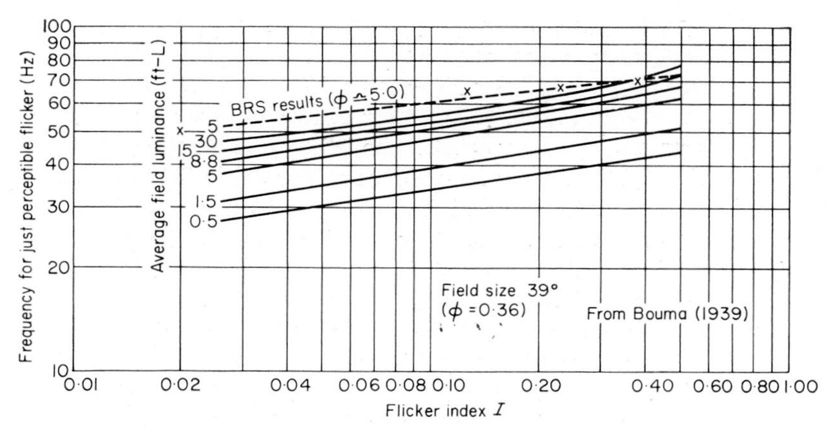

Fig. 5.8 *Relation between flicker index and frequency for just perceptible flicker for different average luminances*
Light-time ratio = 0·5

of the troughs in the waveforms to that of the peaks. This, however, is not so easy where the waveform is of a distorted sinusoidal form as in the case of fluorescent lamps (Fig. 5.6), and to express the visual effect of such waveforms it is necessary to use a method of assessment which corresponds to the way in which fluctuating light is integrated by the visual mechanism. Eastman and Campbell (1952) have put forward a single index for expressing this effect, based on the Bunsen–Roscoe law for describing the effects of intermittent stimulation of the visual mechanism. Fig. 5.7, 5.8, 5.9 show the results of converting different values of modulation and light-time ratio into the Eastman–Campbell Flicker Index in terms of the relationship with CFF or just perceptible flicker.

In order to relate these results to sources used in interior lighting, measurements were made of the light output waveforms for a number of fluorescent

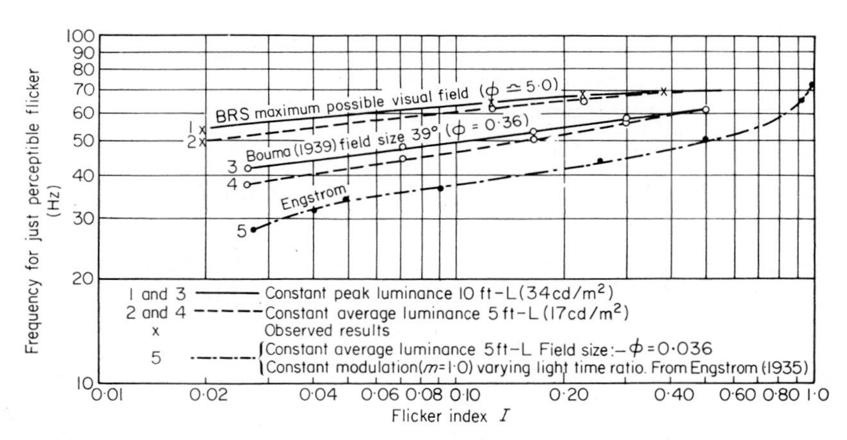

Fig. 5.9 *Relation between flicker index and frequency for just perceptible flicker*

and other lamps, and the Flicker Index determined for these waveforms. The results of these measurements are given in Table 5.2, and it will be seen that the values range from 0·032 for a tungsten filament lamp, through 0·078 to 0·167 for fluorescent lamps, to 0·290 for mercury vapour lamps. Fluorescent lamps at present on the market, intended to give a substantially white light for interior lighting, range in apparent colour of the light emitted from a warm pinkish colour to a cold bluish colour, and the Flicker Index varies considerably from one end to the other of this range. The lamps giving the warmer coloured light at the present time generally make use of fluorescent powders possessing a relatively long afterglow, and it is this afterglow which provides a significant quantity of emitted light during the period in each half cycle when the lamp current is extinguished; this has the effect of reducing the light modulation, and hence the Flicker Index. The lamps giving bluish light (e.g. for colour matching purposes or simulation of north sky daylight) on the other hand, all seem to make use of phosphors which have shorter afterglow times, and hence have a higher Flicker Index; while the plain mercury discharge lamp (MA/V), having no means of emitting light when the current passes through zero, has the highest Flicker Index.

The differing length of afterglow of the different coloured fluorescent powders provides an explanation of the coloured bands commonly seen when a bright object is moved rapidly in a field of illumination from a fluorescent lamp. The reddish or orange light is that remaining as the discharge power drops to zero.

The effect on the perception of flicker of all these characteristics considered in the experimental investigation can be summarised by the formula:

$$F \propto \phi^{0 \cdot 09} B^{0 \cdot 14} I^{n}$$

where F is the critical fusion frequency (or frequency for just perceptible flicker);
- ϕ is the solid angle subtended by the fluctuating field at the eye (steradians);
- B is the average luminance of the field of view;
- I is the Flicker Index of the waveform;
- n is an exponent varying between 0·11 and 0·17 according to the size of the field.

This formula is only known to be valid over a certain range of values of the variables (full details are given in Table III of Collins 1965A), for their interaction one with another is quite extensive. It does, however, indicate the effect of changing the various characteristics.

Perception of 100 Hz Flicker

With a limited extrapolation of the data obtained by the authors and other workers, it is possible to predict the luminances (assumed uniform over the

Table 5.2

WAVEFORM CHARACTERISTICS OF DISCHARGE, INCANDESCENT AND FLUORESCENT LAMPS

Type of lamp	*Power consumed by lamp*	*Current in lamp*	*Mains voltage*	*Ratio Min:Max*	*Per cent. Flicker* $\frac{(Max - Min)}{(Max + Min)} \times 100$	*Ratio Mean:Max*	*Flicker index*	*Rectification**
	Watts	*Amps*	*Volts*					
Tubular fluorescent	80·0	0·832	238·0	0·570	27·4	0·810	0·078	—
'Warm' (pinkish)	80·0	0·888	240·5	0·460	37·0	0·750	0·110	0·014
Tubular fluorescent	80·0	0·846	239·0	0·390	44·0	0·740	0·120	—
'Medium'	80·0	0·870	239·5	0·375	45·5	0·735	0·125	—
	80·0	0·850	239·5	0·355	47·6	0·725	0·130	0·017
Tubular fluorescent	80·0	0·816	236·2	0·253	59·7	0·668	0·153	0·033
	64·0	0·620	210	0·285	55·7	0·695	0·144	0·010
	88·0	0·956	250	0·242	61·0	0·648	0·157	0·097
'Cold'	80·0	0·806	234·5	0·184	68·9	0·668	0·167	0·028
	66·0	0·622	210	0·185	68·7	0·686	0·164	0·012
	89·0	0·942	250	0·174	70·4	0·655	0·173	0·034
200 W Tungsten	—	—	230	0·83	8·7	0·91	0·032	—
125 W MBF/V	—	—	230	0·114	79·5	0·53	0·269	—
125 W MB/V	—	—	230	0·065	87·8	0·50	0·290	—
60 W SO/4	—	—	230	0	100	0·58	0·256	—

* Difference in height of adjacent peaks divided by mean height of peaks.

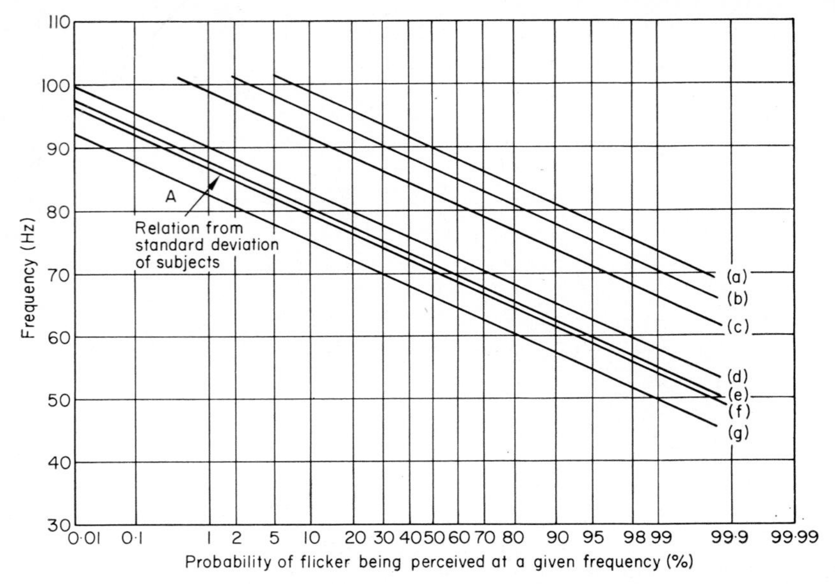

Fig. 5.10 *Relation between probability of flicker perception and frequency for different characteristics of a stimulus presented over the whole field of vision*

The values of flicker index and field-luminance for the above curves are:

Curve	*Flicker index*	*Field luminance* ft–L	cd/m²
a	0·3	50	170
b	0·2	50	170
c	0·1	50	170
d	0·3	10	34
e	0·2	10	34
f	0·38	5	17
g	0·1	10	34

whole visual field) at which 100 Hz fluctuation of the light from various lamps on 50 Hz AC supply would become perceptible.

Taking three lamps of Flicker Indices 0·1, 0·2 and 0·3 (corresponding to fluorescent lamps with long afterglow and short afterglow, and a plain mercury vapour lamp respectively) the luminances of the full field to give just perceptible flicker would be 450, 200 and 100 ft-L (or roughly 1500, 700 and 350 cd/m^2) respectively. These values of luminance are those for average sensitivity of a number of observers, but as they stand they do not give any indication of the probability of flicker being perceived by a few people at field luminances lower than this. To make an estimate of this using the present data, it is necessary to assume that the distribution of the values of frequency for just perceptible flicker shown in histogram form in Fig. 5.4 is a statistically 'normal' one. If this is so, the standard deviation of the results can be calculated, and from this a probability curve (as shown at A in Fig. 5.10) can be drawn. Other probability lines can be drawn parallel to this through the 50% point on the assumption that the form of probability of perception does not vary with field luminance or Flicker Index. These have been drawn in Fig. 5.10 for field luminances of 50 ft-L (or 170 cd/m^2), such as might well occur in a drawing office, and 10 ft-L (or 34 cd/m^2), such as might occur in a school classroom, for each of the three source Flicker Indices; from these curves one can derive the following table of probabilities (Table 5.3):

Table 5.3

PROBABILITY OF FLICKER BEING JUST PERCEPTIBLE AT A FREQUENCY OF 100 Hz

Flicker index of waveform	*Field luminance 50 ft-L (170 cd/m^2)*	*Field luminance 10 ft-L (34 cd/m^2)*
0·1	1 in 150	<1 in 10 000
0·2	1 in 33	<1 in 10 000
0·3	1 in 14	<1 in 10 000

It is evident from the above table that the likelihood of normal 100 Hz flicker being seen in school classrooms lighted only slightly above the Ministry of Education's regulations is very remote indeed, but in drawing offices where illumination levels well above 500 lux (50 lm/ft^2) are not uncommon, there is every possibility that where a large number of workers are concerned, one or two of them may raise complaints of perceptible flicker if precautions are not taken to minimise the extent of the fluctuations.

Subharmonics

In view of the low degrees of probability of perception of flicker under normal interior lighting conditions, it seemed to the authors that the existence of some other factor should be sought in order to be able to account for the small but persistent trickle of complaints brought to their notice. Such a factor was noticed in the examination of the light output waveform records of some of the fluorescent lamps, where it was seen that the two adjacent peaks in the waveform were not of equal height. The effect of this would be to introduce a 50 Hz component into the 100 Hz fundamental flicker frequency, and since 50 Hz fluctuations are much more readily perceptible, it would be expected that a relatively small amount of modulation at this frequency would lead to flicker being perceived.

The 50 Hz component has been regarded as a subharmonic of the frequency of the main fluctuations, and it is apparently due to a partial rectifying action in the lamp discharge (possibly due to asymmetrical emission of the electrodes). Amounts of up to 10% of this subharmonic were found in a sample of old and new lamps examined for waveform.

The effect of small amounts of 50 Hz component on flicker perception was studied on full scale, using fluorescent lamps giving various field luminances, by means of a rectifier in series with the lamp which introduced a small component of the subharmonic according to the adjustment of a shunt resistance across the rectifier. (Details of this study are given in Collins 1956B.) The pattern of the waveform was as shown in Fig. 5.11.

The percentage of 50 Hz component was defined as the difference in height of adjacent waveform peaks in relation to their average height, and seven, six and three subjects respectively made observations at average field luminances of 40, 10 and 4 ft-L (approx. 140, 34 and 14 cd/m^2). All four criteria of discomfort were used, whenever the range was within that obtainable, and Table 5.4 shows the values of 50 Hz component at the above field luminances chosen by the subjects to give the four criteria of flicker discomfort. These results are plotted in Fig. 5.12. It will be seen that at the highest field luminance, very small amounts of the subharmonic can give rise to perceptible flicker, the mean value for six subjects being 2·5% ,with one subject being able to perceive it with 1·2%, and the least sensitive one requiring not more than 4·3% on the average. The amount required is nearly in inverse proportion to the luminance of the field, and rises to between 5 and 26% (average 15·5%) for an average field luminance of 10 ft-L (34 cd/m^2). Levels of field luminance between these two values are quite realistic and even fairly commonplace at the present time, so that it can be expected that some complaints of flicker will be received about an installation in which any of

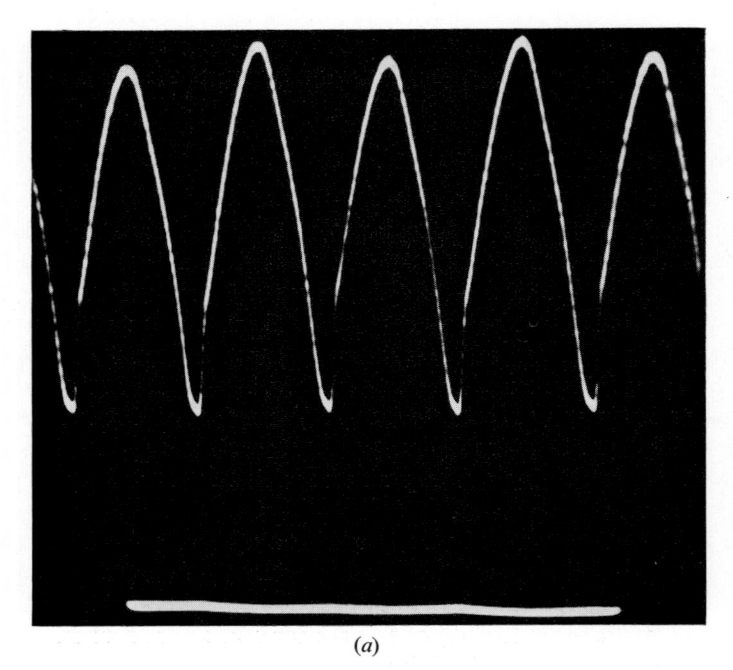

(*a*)

Fig. 5.11 *100 Hz light waveform of 80 W daylight fluorescent lamps with various amount of 50 Hz subharmonic superimposed*

(a) 3% subharmonic
(b) 8·5% subharmonic
(c) 18% subharmonic

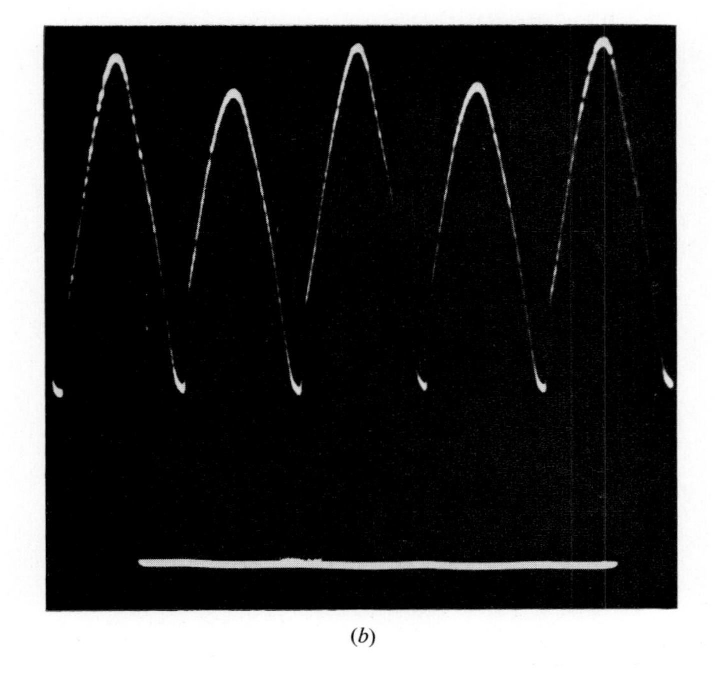

(*b*)

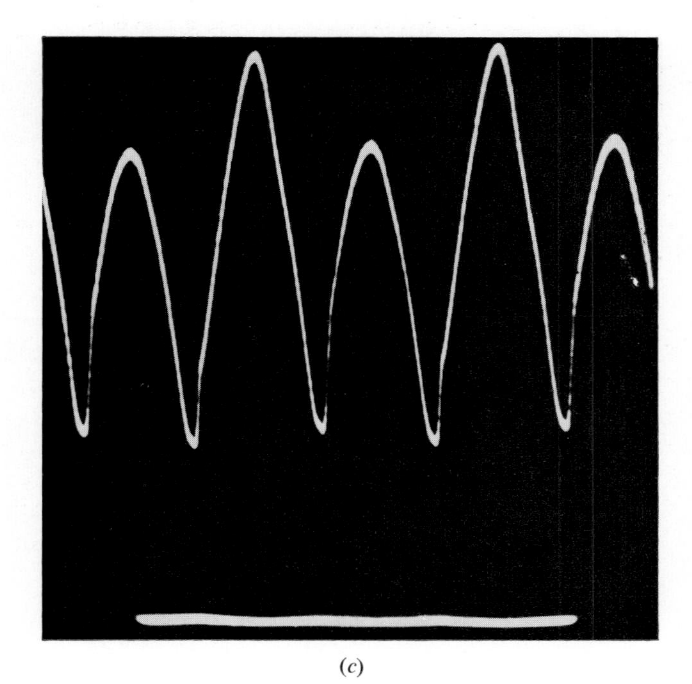

(*c*)

Table 5.4

PERCENTAGE OF 50 Hz COMPONENT REQUIRED IN THE WAVEFORM OF LIGHT OUTPUT FROM A FLUORESCENT LAMP TO PRODUCE FOUR CRITERIA OF FLICKER SENSATION

Average field luminance	*Subject No.*	*No. of observations from each subject*	*Just perceptible flicker*		*Just obvious flicker*		*Just uncomfortable flicker*		*Just intolerable flicker*	
			Mean %	*Range* %	*Mean* %	*Range* %	*Mean* %	*Range* %	*Mean* %	*Range* %
40 ft–L (140 cd/m²)	1	11	1·2	1·6	4·0	3·9	7·4	6·2	14·5	9·9
	2	10	4·3	3·7	9·6	4·5	14·5	3·7	20·6	3·3
	3	6	2·4	2·5	7·4	2·4	13·5	5·5	23·1	5·7
	4	6	1·7	0·7	4·8	4·7	12·7	4·3	20·6	11·3
	5	9	2·2	3·4	8·0	8·1	16·1	6·0	25·7	13·5
	6	11	3·3	2·3	9·2	5·3	19·5	17·9	29·1	14·7
	Average of 6 Subjects	—	2·5		7·2		14·0		22·3	
	[7	6	All < 1·0		All < 2·5		3·2	2·3	5·0	3·0]
10 ft–L (34 cd/m²)	1	8	5·2	6	15·0	16	36·6	35	84·5	32
	6	8	18·2	30	50·6	46	87·4	72	117 or over	
	5	6	16·0	10	39	20	71·5	22	> 126	
	8	6	26·0	5	42	11	74	33	120	20
	9	6	9·0	19	27	29	53	33	105 or over	
	3	6	18·7	17	54	29	73	32	97	50
	Average	—	15·5		38		65·9		> 108	
4 ft–L (14 cd/m²)	5	6	44·5	18	85	34	160	84	> 200	
	6	6	44	32	96	50	200 or over		> 200	
	1	6	18·5	17	43·5	13	69·5	22	110·5	18
	Average	—	35·7		74·8		143 or over		> 170	

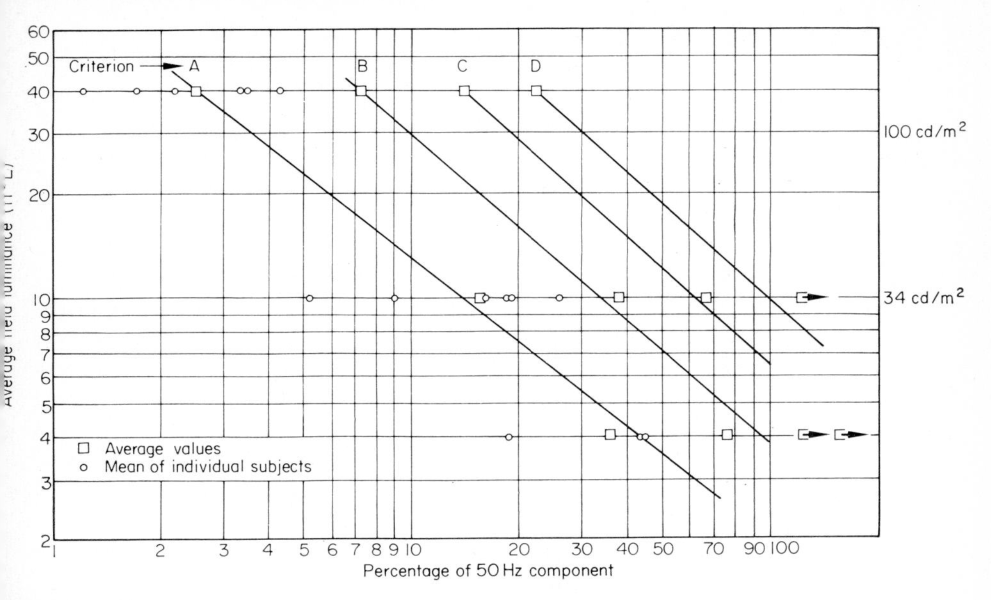

Fig. 5.12 *Relation between field luminance and percentage of a 50 Hz component superimposed on a 100 Hz waveform, for four criteria of flicker sensation*

the lamps exhibit 1 or 2% of 50 Hz component, and quite a few complaints if the lamps exhibit 3% or more.

The relationship between P, the percentage of 50 Hz component, and B, the field luminance for just perceptible flicker, is approximately $P \propto B^{-1 \cdot 5}$, and the form of this law seems to hold for the other criteria of sensation. The steps between these criteria are such that a $2\frac{1}{2}$ to 3 times increase in percentage of 50 Hz component at a given field luminance, or an increase of just over two times in field luminance at a given proportion of 50 Hz component, causes the sensation to change from just perceptible to just obvious flicker. The remaining criteria, however, appear, as far as one can tell from the available data, to be more closely spaced.

The performance of subject number 7 reported in Table 5·4 merits some consideration; it will be seen that he could perceive flicker in a field of luminance 40 ft-L (or 140 cd/m²) with a 50 Hz component of less than 0·1%. This subject was one who was found to be very sensitive in earlier experiments, and under the full-field high-luminance conditions he could often perceive flicker in the visual field when the 100 Hz fluctuations had no subharmonic component added at all. This is a practical illustration that some extrasensitive subjects may be found in any reasonably large group of people, and that special precautions will have to be taken if they are to be able to work under high levels of fluorescent light with complete comfort.

The findings concerning the high sensitivity of the human eye to small components of 50 Hz frequency provide an indication of why the ends of fluorescent lamps can be observed to flicker if they are not obscured by sleeves or the opaque parts of lighting fittings. Measurements of the waveform of light output from narrow sections of a lamp at different distances along its length are shown in Fig. 5.13. It will be seen that at distances of up to 2 inches (50 mm) from the 'end' of the lamp (defined as the junction between the metal cap and the glass of the discharge tube) adjacent peaks of the light output waveform are of considerably different height, thus indicating a large 50 Hz component. This seems to be caused by the occurrence of a dark space near the anode of the discharge at every cycle of mains current alternation. At 3 inches (75 mm) from the end of the lamp, this component has dropped to a very low value, and therefore it is usually only important to screen the first 2 inches (50 mm) of the lamp at either end to eliminate most of the trouble from end-flicker.

Of more concern is the rectification effect discussed earlier in this section; there is nothing which can be done to cure flicker from this cause, if the 50 Hz component is enough to make it visible, except to remove the offending lamp and replace it by another. Engineering staff may be unwilling to deal with complaints of flicker in this way, particularly if the lamp is a fairly new one, or even if it has not reached its allotted span of life before group replacement, but it is important that such lamps should be removed promptly, as the flicker may become very distracting and even distressing after one has been subjected to it for some time.

Having considered how to avoid the conditions most likely to lead to complaints of flicker in a fluorescent lighting installation, we now turn to the precautions which can be taken to lessen the possibility of 100 Hz flicker being visible in an installation producing a large field of high luminance. The only parameter left to vary in this situation is the waveform of the light fluctuations, which is expressed in the basic formula by Flicker Index. The Flicker Index can be reduced by using fluorescent lamps of warmer colour incorporating fluorescent powders with a long afterglow, in preference to lamps of colder ('daylight') colours.

There is a further way of reducing the Flicker Index, which is by trying to fill in the troughs in the light output waveform by adding two or more waveforms of the same frequency but of different phase relationship. The simplest is the standard way offered by several manufacturers of lamp control gear of having two lamps in the same fitting, for example, two 40 W lamps instead of one 80 W lamp, and running one in series with the usual reactive ballast and the other in series with an impedance containing a capacitance. The amount of 'filling in' of the dip in the waveform of light output produced by the standard form of the circuit arrangement (known as a 'lead–lag' circuit) can be seen by comparing Fig. 5.14(a) with Fig. 5.6; it is by no means as complete as might be expected, or possibly as complete as could be obtained by increasing the phase displacement further. There are, however, limitations to the displacement which can be used practically, before incurring losses of power and overcorrection of the circuit power factor. A preferable

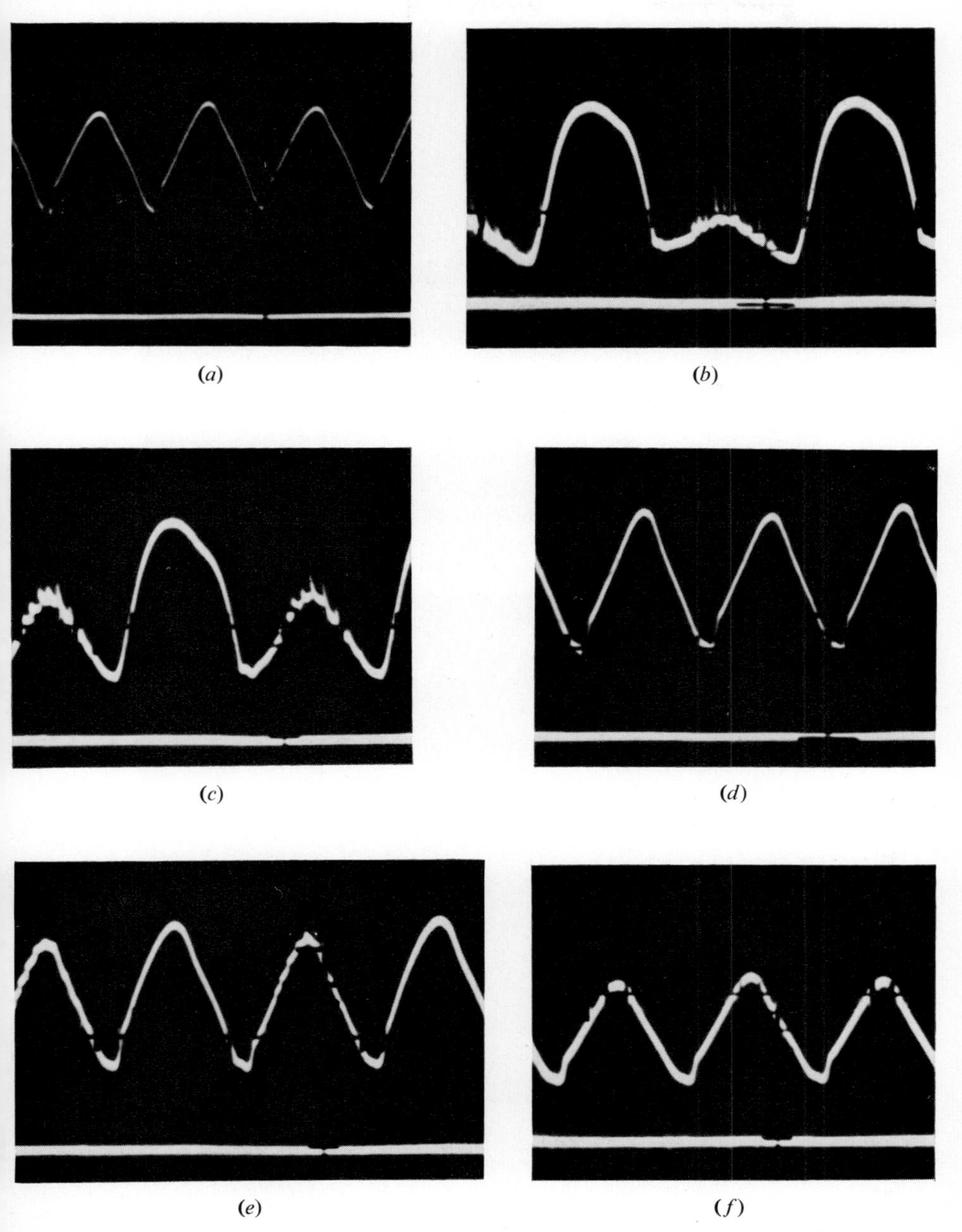

Fig. 5.13 *Light output waveform from different parts of 'warm' colour fluorescent lamp*

(a) Whole tube (d) Centre line
(b) $\frac{1}{2}$ in from end (e) 3 in from end
(c) 2 in from end (f) 6 in from end

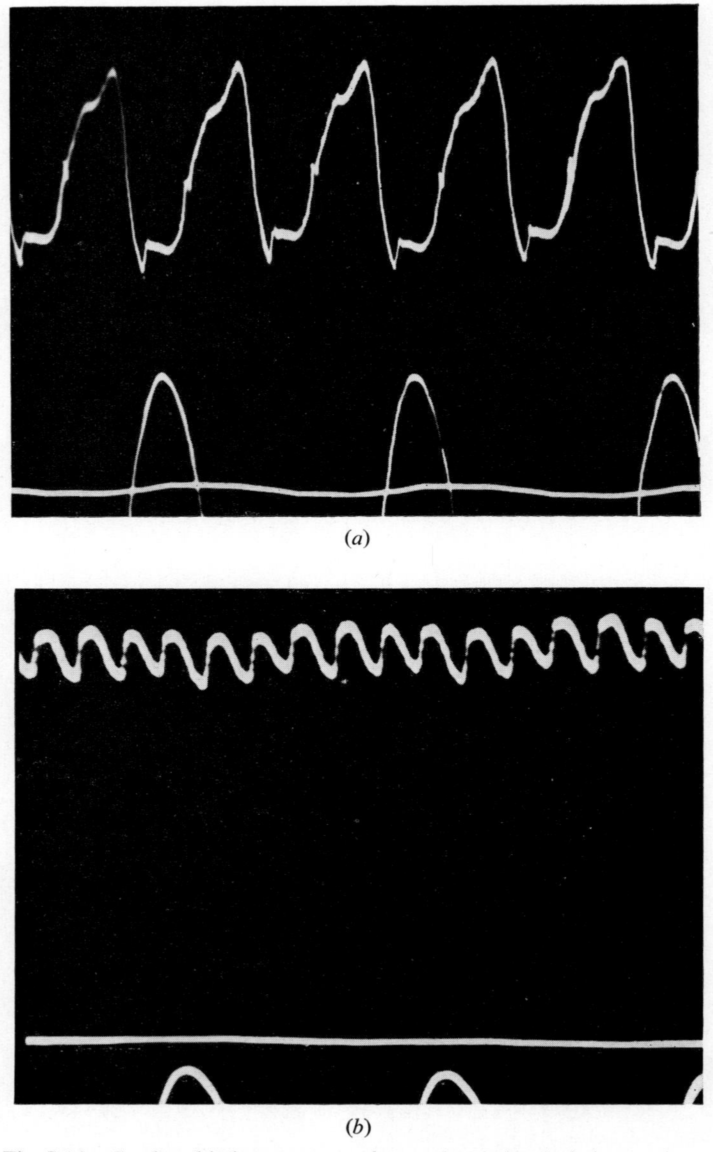

(*a*)

(*b*)

Fig. 5.14 *Combined light output waveforms of MCF/U daylight type lamps*
(a) Two 40 W lamps in a commercially obtainable lead-lag circuit
(b) Three 80 W lamps fed from 3 different phases of 50 Hz 3-phase supply

way of achieving phase displacement between the current in the discharges of different lamps is to feed them from different phases of a multiphase supply; for example, have three adjacent lamps in an installation (in the same fitting if possible) fed from the three different phases of a three-phase supply. This type of phase displacement has the effect of trebling the supply frequency, and this is shown in Fig. 5.14(b). It will be seen that the amplitude of the fluctuations from the combined light sources is very much reduced by virtue of the increase in the frequency. Fluorescent lighting arranged on a three-phase supply is virtually free of visible flicker.

The reduction in amplitude of fluctuation as the effective frequency of the light output waveform increases would seem to indicate the desirability of running fluorescent lighting on a high-frequency supply. This is, in fact, very advantageous, both from the point of view of reduction of the possibility of the perception of flicker at supply frequency (subharmonic) and light fluctuation frequency (twice the supply frequency), and there is also a considerable economy to be effected because much smaller ballast gear can be used. Some use of high-frequency supplies for fluorescent lighting is being made in buildings in the United States, while in Great Britain and many other parts of the world, high-frequency supplies are common for fluorescent lighting in aircraft and vehicles. There is a certain amount of technique developed for fluorescent lighting at high frequencies, and it is not unreasonable to predict that in the future more consideration is likely to be given to high-frequency fluorescent lighting in buildings. The most likely application is in hospitals where very large and more compact buildings are to be expected, with some of the treatment and service rooms provided with a high level of artificial illumination to compensate for the lack of natural daylight. Large office blocks and engineering factory drawing offices are other buildings which are sufficiently compact, or which have easy enough access to services, to make it worth while installaing a special high-frequency supply for the fluorescent lighting. Developments of this kind await the manufacture of inexpensive control gear. At the moment, special high-frequency control gear is expensive because it is nonstandard. It is to be hoped that before long a way of breaking the deadlock will be found following the spread of solid-state control devices, and that realistic cost comparison studies will show the extent of the economies which can be effected.

Stroboscopic Effects

The advantages of high-frequency fluorescent lighting are not, of course, confined to the perception of direct flicker. The stroboscopic effects of fluctuating light sources which depend upon the nature and speed of movement of the object observed, its contrast with the background, as well as with the frequency of the illuminant, would also disappear completely under high-frequency fluorescent lighting. Stroboscopic effects are negligible under filament lighting at normal frequencies, and are less when fluorescent lamps with long afterglow (usually of the 'warm' type) are used. Split-phase circuits, such as the 'lead–lag' circuit, are often adequate to eliminate annoying effects. There seems some evidence from this that, provided the Flicker Index is below 0·08 (at a supply of 50 Hz), stroboscopic effects will cause little trouble.

References

Bouma, P. J. (1939): Work presented by W. Uyterhoeven. 'Periodic Variations of the Light Output of Gaseous Discharge Lamps.' *Proceedings CIE Tenth Session*, **2**, 120–128.

Collins, J. B. and R. G. Hopkinson (1954): 'Flicker Discomfort in the Lighting of Buildings.' *Trans. Illum. Engng. Soc.* (*London*), **19**, 135–158.

Collins, J. B. (1956A): 'The Influence of Characteristics of a Fluctuating Visual Stimulus on Flicker Sensation.' *Ophthalmologica*, **131**, 83–104.

Collins, J. B. (1956B): 'The Role of a Subharmonic on the Wave Form of Light from a Fluorescent Lamp in Causing Complaints of Flicker.' *Ophthalmologica*, **131**, 377–387.

Collins, J. B. and R. G. Hopkinson (1957): 'Intermittent Light Stimulation and Flicker Sensation.' *Ergonomics*, **1**, (1), (November), 61–76.

Eastman, A. A. and J. H. Campbell (1952): 'Stroboscopic and Flicker Effects from Fluorescent Lamps.' *Illum. Engng.* (*New York*), **47**, 27–35.

Engstrom, E. W. (1935): 'A Study of Television Image Characteristics.' *Proc. Inst. Radio Engrs.*, Part II, **23**, 295–310.

Ireland, F. H. (1950): 'A Comparison of Critical Flicker Frequencies under Conditions of Monocular and Binocular Stimulation.' *J. Exper. Psychol.*, **40**, 282–286.

Lythgoe, R. J. and K. Tansley (1929): 'The Adaptation of the Eye: Its Relation to the Critical Frequency of Flicker.' Medical Research Council Special Report Series No. 134. HMSO, London.

Porter, T. C. (1902): 'Contributions to the Study of Flicker (II).' *Proc. Roy. Soc.*, **A, 70,** 313–329.

6

Visual Fatigue

Bad lighting is well known to lead to a form of fatigue which the layman calls 'eye strain' or 'tired eyes'. No precise clinical definition has so far been found to describe this condition. The term 'visual fatigue' is often used but there is no common agreement that visual fatigue really exists, or, if it does exist, that it can be measured by any objective means. Bartlett (1953) and others following him have defined fatigue as a deterioration in an activity as a direct result of being engaged in it. This definition immediately presents further difficulty for those studying visual fatigue, since subjective symptoms of fatigue generally appear in those engaged on visual tasks long before deterioration in the performance of the task appears.

The incentive for the study of the measurement of visual fatigue would largely be lost if a technique could be developed demonstrating a reliable and repeatable relationship between the output of any type of visual work and the quantity or quality of the lighting under which it is performed. However, in spite of many recent claims to have shown increases in productivity corresponding to increases in illumination in working areas, we are still far from avoiding the need for another method of measuring the effects of lighting on those working under it. There are still a great many people whose work does not directly produce a tangible, measurable output; it may well be that these people are in the majority. Shop assistants, vehicle drivers, railway signalmen, teachers and nurses do not have the task of turning out tangible items of work of which the quantity and quality can be measured in relation to the environmental design. All would admit, however, that the conditions of the visual environment under which they carry out their work have a considerable effect on their wellbeing, and ultimately on the performance of their duties.

The performance of a visual task as distinct from casual seeing necessarily involves some effort on the part of the individual. This effort may be muscular, bringing into play the accommodation and oculomotor muscles in reading, in searching for detail or in watching for a change of pattern. From all we know of the visual process, it is certain that this muscular effort is made more difficult by bad lighting either in the form of insufficient luminance or undesirable luminance pattern in the visual field. Conversely, the effort may be reduced by producing optimum conditions of luminance and luminance distribution. Even when only casual seeing is involved, an unsatisfactory visual environment is an 'emotional affront' (Weston 1953) which may well cause a form of fatigue. Human adaptability to working under a very wide range of conditions is so great that unless conditions are very bad, no effects of the effort involved in the visual task are usually apparent after short periods of work. They do, however, make themselves felt after long periods at the task and the individual may then complain not only of 'eye strain' but of generally feeling tired, less alert, and these symptoms are regarded as those of visual fatigue or perhaps more precisely of 'fatigue visually occasioned'.

The precise origins of visual fatigue or of general fatigue visually occasioned are not easy to find. It may be of value to consider the whole visual process in the course of this search. The part of the retina of the eye with the highest discrimination of detail and contrast at normal (photopic) levels of luminance is the fovea centralis occupying an area subtending about 2° in diameter in the centre of the visual field, and the first essential in performing a visual task is for the eye to direct itself to bring the most critical part of the task on to this central part of the retina. This direction is performed partly by the muscles of the head and trunk, which perform the coarse control necessary, and partly by the extrinsic oculomotor muscles which perform the fine adjustment of movement (other things being equal, the head will be held so that the minimum amount of movement of the eyes themselves from their central 'straight ahead' position is called for).

These extrinsic oculomotor muscles do not, in fact, hold the eye rigidly directed but, as shown by, for example, Lord and Wright (1950), small involuntary movements are continually imposed on the eyeball, the significance of which in the visual process has been demonstrated by Ditchburn and his colleagues (1952). As long ago as 1932 Lancaster showed that the more unfavourable the seeing conditions, the more critical become the fixation movements. However, as far as the voluntary muscles are concerned, their function is to keep the task as nearly as possible centred on the fovea whether the task is a fixed one such as a small visual display to be watched, or one involving movement such as following a moving object, scanning a scene, or reading a printed page. Lion (1952) has found that they do not show any decrease in power after prolonged activity, even though the subject is very tired.

Inside the eye itself the ciliary muscles have the function of focusing the image on to the retina by flattening the lens or allowing it to bulge, thus altering its focal length to accommodate for different distances of objects from the eye. The iridomotor muscles (entirely involuntary) adjust the amount of light allowed to fall on the retina. These muscles control the aperture of the pupil which, in addition to limiting the illumination on the retina, also results in an improvement in the definition of the image as well as increasing the depth of focus.

When light reaches the retina, it is converted by a complicated electro-chemical process into coded electrical information which is transmitted via the optic nerve to what are known as the higher visual centres of the brain. Here the most elaborate processes are performed: those of the interpretation of the information received. Some of this information is processed and fed back via the autonomic nervous system to control the focusing and pupil adjustments of the eye, while some of the information is passed to the cerebral cortex for decision making, storage, etc.

In view of the fact that performance of a visual task involves a certain amount of muscular activity, some types of fatigue visually occasioned will be associated with muscular fatigue, while other aspects will be manifest in general or central fatigue, or in tiredness which cannot be associated necessarily with muscular activity. The former type of effect is the only one to which the term 'eye strain' can strictly be applied and moreover is the only one which can directly cause any sensation of fatigue or discomfort in the eye itself. Prolonged performance of a visual task or a task made difficult by poor lighting may well lead to the well known symptoms of soreness, irritation and general discomfort of the eyes, but unless muscular fatigue is involved, these symptoms will be manifestations, probably among others such as headaches, of central fatigue or general tiredness.

In the early 1930s concern was felt over the possible production of eye strain by viewing cinema pictures in much the same way as more recently there has been concern about the viewing of television. Snell (1933) analysed very clearly the part played by the several components in the visual process. Weston (1953, 1962) and Carmichael and Dearborn (1947) have described in some detail the operation of the various muscles involved in a visual task, and Weston has also discussed whether the retina itself can show any sort of fatigue effect. Snell believed that it could, and demonstrated the reduction in critical fusion frequency caused by working in flickering light. This reduction in CFF is, however, now considered to be a case of adaptation rather than fatigue, and in spite of the care taken by Snell to distinguish between adaptation and fatigue, it is believed that his experimental work was not adequate in this respect. One thing which is quite clear is that the retina itself cannot experience any sensation other than that of light, nor can the optic nerve fibres convey any messages other than in terms of light or to other centres than the visual centres. If the retina or optic nerve become fatigued in any sense at all, it can only be in the sense that their input to the visual centres varies in quantity or quality and not in any sense that we can directly feel.

It will be evident that it is usually a misconception to speak of 'eye strain' being caused by bad lighting since it is only indirectly that defects in lighting a task can lead to muscular discomfort. Such defects do occasionally occur, and it can happen that the head or upper part of the body cannot assume a relaxed position because of the necessity to avoid casting shadows on the work or of avoiding bright reflections in the work itself or bright sources or reflections of sources close to the work. Insufficient illumination of the task can also lead to muscular fatigue or tiredness if it compels the worker to bring his eyes nearer to the task in order to increase its apparent size, that is, the size of the image on the retina, and so improve the discrimination of detail or of low contrast. Even if the task is brought nearer to the worker instead of the other way round, there is still a possibility of muscular fatigue being felt in the ciliary muscle if this is forced to apply a large amount of accommodation to focus the lens for short distances for long periods of time without relaxation.

Much used to be said, in the days before those concerned with lighting practice linked their work with that of those studying the physiology of vision, about the harm to vision caused by bad lighting. It will, however, be clear from the discussions so far that apart from the effects of prolonged adoption of a bad posture on growing children, none of the effects of bad lighting can cause more than a temporary discomfort in the eye itself. The lens of the eye will not permanently assume a high degree of curvature, or a ciliary muscle weaken due to the continuous use of large amounts of accommodation.* The retina cannot suffer permanent damage because effort is being made to resolve detail beyond its limits of discrimination.

The avoidance of direct muscular fatigue or 'eye strain' therefore requires the exercise of only the most elementary rules of illuminating engineering to ensure that a comfortable posture and focusing distance are easy to attain. Lighting must be designed so that there is not the need to avoid shadows or to dodge bright reflections. There should be adequate illumination and there should be a comfortable background provided on which the eye can occasionally rest and relax its accommodation.

The aspect of the problem which is of greater concern to those who are trying to develop techniques for the design of the best possible visual environment is the fatigue and tiredness which results from prolonged work under poorly designed lighting where the visibility of the task is not satisfactory or where there are distractions in the field of view. This fatigue may indeed only be noticed when compared with good conditions. People working in unsatisfactory conditions are not likely to complain unless symptoms are severe.

It has been shown that direct connections exist between the retina and the autonomic nervous system (Le Grand 1961) and that the amount or the rhythm of cyclic change of environment can effect the disposition to work (Schneider 1963). It is, however, reasonable to assume that the influence of the quality of the lighting on the moods or 'affective states' of those working under it must be through the higher centres of the nervous system, and that the methods of examination which tell us most about the state of the central nervous system are most likely to be successful in showing the effects of lighting quality.

* Bronner (1967) has suggested, however, that myopia might be caused in growing children by elongation of the anterio-posterior axis of the eye resulting from continuous efforts of convergence.

The many methods which have been examined by different workers in attempting to find a function which would reflect differences in the central nervous system caused by differences in the quality of lighting can each be placed in one of two categories:

1. Those methods involving a function strictly of vision ('visual functions').
2. Those methods involving a measurable function of the eye or some other organ or system of the body ('physiological functions').

The first category includes measurements of critical flicker fusion frequency (CFF), the speed of contrast discrimination, performance of visual tasks, and possibly the electro-optical sensitivity. The second category includes muscle tone and muscle action potentials, finger tremor, equilibrium movements, accommodation reserve and speed, pupil diameter measurements, eye movements, convergence control, blink rate, upper frequency audibility threshold, visual-motor reaction time, and manual dexterity. Michael (1954) has discussed many of these tests for visual fatigue, and Grandjean (1959) has tested many of them in the course of a study of the working conditions and performance of telephonists.

The tests involving visual functions will be considered first:

Speed of Contrast Discrimination

Baumgardt and Le Grand (1956) refer to contrast discrimination tests as the 'glare test'. The test was used by Ségal (1950) who acknowledged the earlier work of Lioublina (1944) in demonstrating the use of the Troukhanov visibility meter as a measure of previous visual activity. In essence the test consists of a measurement of the time taken to recognise the orientation of a Landolt ring when its contrast is reduced by veiling brightness. Ségal obtained positive results with this test, in that when measuring the perception time (which he called adaptation time) of people who had been reading for an eight-hour day, he found that this time was significantly longer after they had been reading under fluorescent lighting fed from a single phase supply than when the lighting had been from lamps distributed between three phases (see Chapter 5). This result was in line with his previous work showing the increased activity of the cortex with greater amplitude of fluctuation of the light stimulus, and might therefore be taken as evidence that the method does provide an indication of fatigue of the central nervous system. Baumgardt and Le Grand (1956), were however, unable to repeat this result. This does not necessarily mean that the method is invalid, and in fact they admitted that they were not able to duplicate Ségal's adaptation conditions before and after working, and this is probably an important factor in the situation. This lack of correspondence between investigations by two independent teams suggests, however, that further study of the glare test method is required before it can be recommended as a means for assessing a form of visual fatigue. Brozek and Simonson (1952) used a similar test which they described as 'recognition time for stimuli (dots) of threshold size determined by means of a stopwatch'. The increase of this time after the subject had performed an exacting visual task for two hours was found to vary with illumination level, and was found to be reduced significantly when the illumination on the task was increased from 5 to 300 lm/ft^2 (50 to 3000 lux).

Critical Flicker Fusion Frequency

The use of the CFF as an indicator of fatigue has been studied extensively, especially over the last twenty years. Brozek and Keys (1944), Brozek *et al.* (1950), Simonson with various colleagues (1941, 1948, 1952), Snell (1933), and many others obtained rather conflicting results. Ryan, Bitterman and Cottrell (1953) concluded that the changes in CFF did not reflect precisely the visual strain in the situation but rather the general difficulty, stress, alertness and concentration required by the task. More recently Grandjean (1959) with his colleagues (1953, 1955 and 1960), and Werner and Grandjean (1960) have established the usefulness of the measurement of the reduction of CFF for revealing the demands on the eye in industrial working places and laboratories. They have shown (1955) that exhausting physical work or quite difficult reading work lasting one hour did not lower the CFF, but that counting dust particles under a microscope for one and a half hours, reading with only one eye for one hour, a day's work radiographing ampoules, telephone operating or needle finishing, which are all very difficult tasks requiring visual concentration, reduce the CFF significantly. Moreover, the reductions in CFF could be lessened by resting during the midday break or by improving the lighting (by reducing contrasts at the edges of the field of view), while the use of fluorescent lamps without dephasing markedly amplifies the CFF reduction.

These results are in good agreement with those obtained by Simonson and Enzer (1941) who measured the CFF for a number of hospital staff engaged on tasks which, though physically easy, were nevertheless quite fatiguing. They found that reduction in CFF at the end of the day more or less paralleled the expressions of subjective fatigue.

Grandjean and Perret (1961) noted that while a number of other workers had also found reductions in CFF after different types of work, many others (admittedly not studying tasks which were of very great visual difficulty) had not found appreciable changes. They therefore investigated the effect of certain conditions of measurement upon the results obtained and arrived at four important conclusions:

1. If the flickering source was viewed through an artificial pupil of 1·8 mm diameter, the actual values of CFF were altered but there was no significant change in the reduction in CFF measured after the task. They interpreted this result as indicating that reductions in CFF as normally

observed are not due to the action of fatigue causing a reduction in pupil aperture through parasympatheticotony.

2. Increasing the time taken for making the measurement of CFF (from 10 to 16 seconds) lowered the value of CFF and partially masked the 'fatigue' drop. This finding is obviously of the greatest significance in working out an experimental technique for making CFF measurements.
3. Interrupting the intermittent light between two measures of CFF to shorten the total time of exposure did not alter significantly the values obtained.
4. Values of CFF measured under identical conditions (including, presumably, degree of fatigue) on two consecutive days showed a significant increase on the second of the two occasions, which these authors interpreted as evidence of a practice effect. (Strangely enough, they did not continue to repeat the measurements to see if the practice effect levelled off eventually).

Performance Tests

Performance tests for the indication of visual fatigue are usually measurements of speed or accuracy in performing a task which involves quick and careful seeing. Visual performance tests can, where the task is suitable, be measures of the performance of the normal work itself, or they can be administered separately, for example, interpolated in the course of the normal work. If the task is such that measures of output in quantity, speed, accuracy or variability can be made, it is possible to overcome the objection which is often heard that interrupting the task to take a measurement or stopping the task at the end of the period of work automatically provides a condition of recovery. The intention is, of course, to measure the extent to which the subject is fatigued at the end of his period of actual working.

Bartlett (in a private communication to the authors) pointed out that fatigue effects may occur in unknown phase relationships with the performing of the task and that any interpolated tests or periods in which the performance of the task is measured should ideally be distributed at random intervals throughout the period of the test.

There are a great many different forms of performance test. Grandjean (1959) used a rather unusual test of performance in which he used mental addition as a means to estimate fatigue. In fact, he found no significant change in either speed or number of mistakes made by telephone operators after their work shift as compared with their performance before it. Other tests (CFF reduction, see above) had revealed significant changes, so he concluded that the mental addition test was of no value.

Weston (1945) devised a test of visual performance (see Chapter 3) in which the subject is presented with a sheet containing rows and columns of broken Landolt rings with the gaps occurring at one of eight possible orientations. He is then required to cross out all the rings which have the gap in one specified orientation. The time taken to perform this operation is measured, and is weighted according to the errors made. If desired, an additional correction can be made to allow for the time required to perform the actual mechanical act of marking. The visual difficulty involved in this performance test can be altered by the use of different charts with different sizes of the critical visual element (the gap in the ring) or different degrees of contrast. Saldanha (1955) tried this test before and after a two-hour period of working on a vernier gauge, but found no change in performance by subjects who were expressing feelings of fatigue.

A very similar test to that of Weston has been used by Grandjean (1959) which he refers to as the Bourdon test. Letters are used in a meaningless order instead of Landolt rings, and in the course of the test one specified letter has to be cancelled by the subject whenever it is encountered on the chart. Grandjean found, however, that contrary to expectation the mean value of the time taken by a group of telephone operators to perform this test was significantly *reduced* after their work shift. He interpreted this as 'a consequence of a decrease in cortical inhibitions in a state of fatigue'. The results of these three independent investigations suggest that interpolated performance tests do not offer much promise as measures of fatigue produced by a visual task and that the view that 'a change is as good as a rest' after performing a long visual task may, in fact, be correct.

Two examples may be noted of the second category of performance test, that is, those which can be used as the fatiguing task and scored continuously or at intervals over a period of time. One is the test used by Brozek and Simonson (1952), and the other, that developed by Saldanha (1955).

One difficulty inherent in measuring output as an indication of performance over a period of time is that the quality of output at any given moment is to a large extent a function of motivation at that time. It is frequently found that where the activity is largely mental, no decrease in quantity of output occurs, although the subject may express definite feelings of fatigue. For this reason the incidence of errors in the performance of a task is found to be a better measure of the effects of prolonging the task.

Brozek and Simonson and Saldanha have both used a measure of the errors in performance rather than the quantity of output. The former set their subjects viewing small letters passing behind a narrow slit at irregular intervals and irregular heights, visible only for a fraction of a second. These letters they had to copy down for a period of two hours and the performance at the beginning and end of the work period was scored by noting how many letters were correctly copied during a period of six minutes. The task was performed under a range of illumination levels between 2 and 300 lm/ft^2 (20 to 3000 lux) and it was found that:

1. There was a significant increase in average performance up to 100 lm/ft^2 (1000 lux) and therefrom a slight fall off as the illumination was increased to 300 lm/ft^2 (3000 lux).

2. The decrement in performance (comparing that at the beginning with that at the end) showed a similar (numerically inverse) trend, that is, an optimum value at 100 lm/ft^2 (1000 lux) (minimum decrement).

The task used by Saldanha was devised in collaboration with the present authors to develop a test which would be related to the visual activity involved in draughtsmanship. The subject was required to set a fine vernier gauge to a series of given values. The gauge could be read to 0·001 in and the accuracy with which it was set could be measured to 0·0001 in. The subject paced himself in that after making a setting, he pressed a switch which caused a new number to be presented to him for him to set on the vernier. This switch also caused a camera to record the accuracy of his settings for the experimenter, but the subject himself was not aware of the quality of his performance. He was told that both accuracy and speed would be scored and that he was to continue with the task until told to stop. The task was, in fact, continued for two hours without a break and the errors were averaged over every

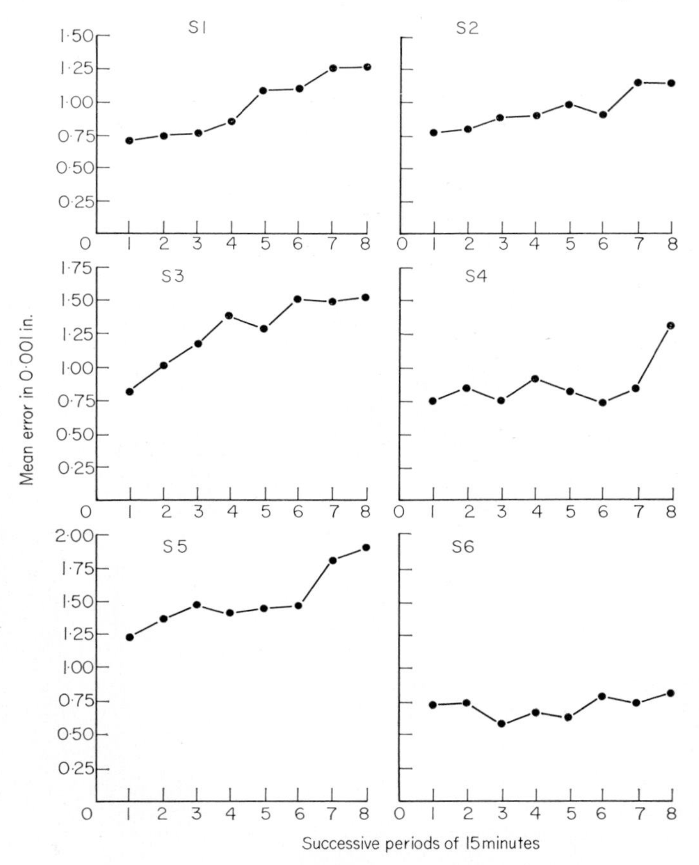

Fig. 6.1 *Variation in error with time*
(Saldanha's experiment, individual subjects)

quarter of an hour. Although the progress of the errors for individual subjects exhibited a certain amount of variability, nevertheless when the average error for a group of eighteen subjects was plotted, a progressive increase over the two hours was noted (see Fig. 6.1). This increase in errors seemed to correlate with the fact that most subjects found the task very tiring.

Saldanha was mainly concerned with the effect of such things as whether knowledge of results or taking rest pauses affected the performance, and his task was well suited to this when carried out by a team of subjects whom he had selected and trained for the purpose. The task seemed also very suitable as a means for testing the effect of illumination level and luminance distribution in an interior. The present authors therefore adapted Saldanha's test for a detailed study of the effect of illumination and luminance distribution on visual fatigue when performing difficult visual tasks such as those which might be encountered in an engineering workshop or a drawing office.

Saldanha's apparatus was duplicated at the Building Research Station (Collins 1959), using a vernier gauge measuring to thousandths of an inch, mounted on a rigid base and connected to a nut working on a lead screw so that settings of the vernier scale could be accurately made by rotating a hand wheel attached to the lead screw (see Fig. 6.2). Spring pressure on the nut eliminated backlash. Bearing on the vernier slide was the feeler of a sensitive dial gauge reading to one ten-thousandth of an inch, the dial of which was mounted facing inwards to a recording cabinet not visible to the subject himself. The numbers to be set by the subject (in inches and thou-

Fig. 6.2 *BRS vernier gauge setting apparatus*

sandths) were chosen at random and were typed on to both sides of a loop of 35 mm perforated photographic paper. This loop travelled round inside the recording cabinet, being driven by a solenoid and ratchet mechanism between every reading, and the number to be set appeared in a window in the cabinet just above the vernier scale. An instrumentation camera photographed the number on the back of the paper opposite the window in the cabinet, together with the dial gauge reading and a stopwatch. The activation of the camera was controlled by the subject, for on operating the switch to make the next figure appear, the photograph was automatically taken during the delay period before the number changed (Fig. 6.3). Illumination of the vernier gauge required special treatment. The most important feature of the lighting system was to ensure a light ceiling and upper part of the wall behind the subject in order to secure adequate contrast of the vernier engravings with the polished surface of the scale. The area surrounding the vernier scale was made as light as possible in a deliberate attempt to produce a visual pattern which would be expected from previous experience to be found tiring.

Subjects received initial training to enable them to set to the required accuracy (0·005 in). The subjects were chosen from staff associated with lighting research, with the addition of one mathematician and two trained mechanics.

The illumination on the vernier scale was set at 14 lm/ft^2 (150 lux) for the first series of tests, and the average errors over quarter-hourly periods for the duration of the two-hour run were examined. It was found that a very slight increase over the whole two-hour period occurred in the average results of the first batch of subjects, but that individual results were variable, as many subjects showing a decrease in error as showing an increase. In order to obtain a more definite tendency, tests were repeated under a low illumination level (1 lm/ft^2 or 11 lux), but again no consistent pattern of increase in error or of increase in variance of error was found. (It was interesting that the general level of accuracy of the trained mechanics was worse than that of the research workers.)

One other attempt to influence the performance of the subjects was made; a stress was applied after the first half-hour by telling them that only settings with an error of not more than 0·5 thousandths of an inch would be counted, and telling them after each setting whether it would be accepted. All errors for the subjects tested in this way were plotted and analysed, and these showed that while one of these subjects showed a marked decrease in performance and increase in variance under this form of stress, others were not affected.

A final attempt was made to check whether the apparatus or technique was in some way so different from that of Saldanha to account for the lack of correspondence with the results which he had reported. The equipment was taken to Saldanha's laboratory and four of the BRS-trained subjects were tested under his direction and adopting his full procedure. Little or no decrement in performance was found, and certainly no correspondence with the result previously obtained by Saldanha, either in the analysis of error or variance of error.

No explanation of this lack of correspondence has so far been found.

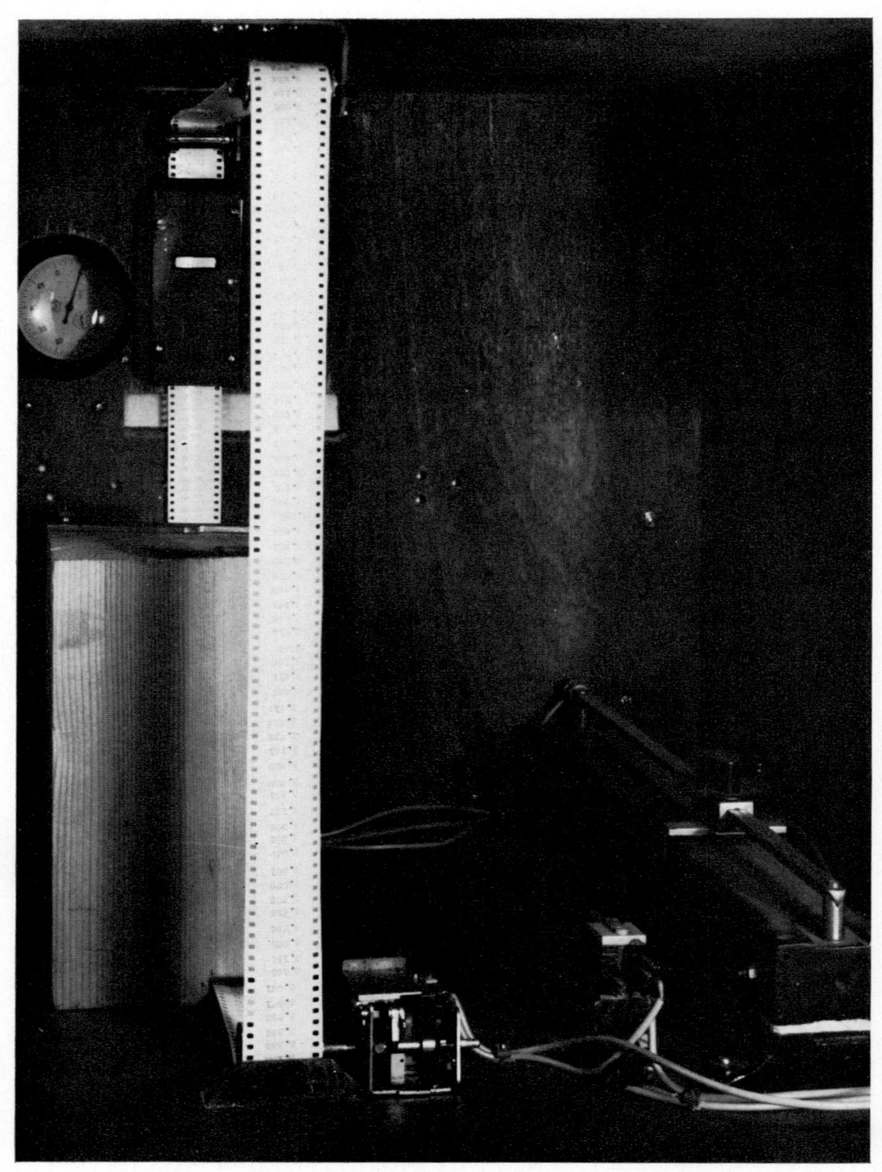

Fig. 6.3 *Operating mechanism and recording position of BRS vernier gauge apparatus (camera removed)*

Suggestions have been made that motivation and training influence the situation, but this obviously is not the complete explanation. The unfortunate conclusion could not be avoided that this test, which initially offered high prospects of serving as a useful objective measure of visual fatigue, could not, in fact, be considered satisfactory for this purpose.

The unreliability of performance itself as a useful measure has been confirmed by Khek and Krivohlavy (1968), but they found that variability of performance (reading Landolt rings) was a more sensitive indicator of the effect of illumination level and glare on the worker.

The conclusion of all this work on performance tests is inescapable; though some tests do undoubtedly show positive results in the hands of some workers with some groups of subjects, none of the tests so far published can justify a claim for complete objectivity. Certainly it would be unwise to base any major conclusions about lighting design or illumination level on the results of any of these experiments.

Electro-optical Sensitivity

Motokawa and Suzuki (1948), in studying the phenomena of phosphenes (sensations of light produced by the direct stimulation of the nerve endings or nervous system) found that a flickering sensation of light could be produced with the eyes closed by applying intermittently a low potential difference (less than 2 V) across silver electrodes on the skin near the eye. The differential between the threshold voltages for just perceptible sensation to be produced by increasing stimulus voltage, and for the sensation to become just imperceptible again on lowering the stimulus voltage, was found to rise appreciably during the working day. It was also found to fall during rest periods. For telephone operators this threshold difference was found to follow closely the curve rate of calls handled and was therefore presumed to be a good indication of fatigue related to the rate of working. This method of measuring fatigue has disadvantages as it requires rather careful experimental techniques, and it is said to need well-trained subjects. It does not seem likely, therefore, to find wide application for field studies. One possible variant has, however, been suggested by Ségal (in a private communication to the authors) in connection with the measurement of the increased time constant of the nervous elements due to prolonged activity. He has suggested that the discrimination interval threshold between two consecutive stimuli should show an increase with fatigue, and that the visual nervous centres might be stimulated to produce phosphenes by two consecutive electrical impulses. The threshold time interval between the two impulses, at which they only just merge to produce a single sensation, should be measured before and after a prolonged visual task. No developments have so far followed on Ségal's proposals.

All the tests so far considered have involved strictly visual functions. Tests involving physiological functions are now considered:

Muscle Action Potential and Tension

The measurement of indirect physiological functions as a means of assessing visual fatigue, or fatigue visually occasioned, has a long but not altogether satisfactory history. Luckiesh and Moss over a long period undertook different experiments to try to demonstrate the value of better lighting in improving vision. In one study (1933) they recorded the pressure unconsciously exerted by the fingertips while reading, and showed that this pressure decreased considerably as the illumination level was raised from 1 through 10 to 100 lm/ft^2 (10 through 100 to 1000 lux). They also demonstrated that fingertip pressure doubled when a glaring source was introduced into the field of view.

Another method of assessing tension used by Luckiesh and Moss was to measure the heart rate electrically. They found the heart rate decreased by ten per cent after a one-hour period of reading under an illumination of 1 lm/ft^2 (10 lux) but by only two per cent after reading the same task under 100 lm/ft^2 (1000 lux).

Pupillary diameter has also been suggested as an indirect physiological measure of visual fatigue. Michal records that Luckiesh and Moss measured pupil diameter in a number of subjects before and after work and found an average increase in pupil diameter of 0·5 mm. Luckiesh himself (1944) was not as enthusiastic about the value of pupillary diameter measurement as he was about the other measurements such as heart rate and finger pressure. Michal (1954) also criticised the conclusions from the Luckiesh and Moss experiment and reported more favourably on the pupillography of Löwenstein and Löwenfeld (1951, 1952), who recorded the change of pupil reaction time to light stimulus at different degrees of fatigue. They reached the conclusions that fatigue caused by heavy manual work had the same effects as repeated light irritation, and that since the vegetative nervous system is affected, the total nervous fatigue is of central origin, and that the sympathetic nervous centres are tired sooner than the parasympathetic centres, and that cortical centres are fatigued sooner than the subcortical centres.

Eye Movements

Brozek and Simonson (1952) included measurements of the maximum rate of voluntary eye movement by recording the number of fixation cycles of amplitude about 15° made by subjects in a period of 10 seconds when they were asked to sweep the eyes backwards and forwards between two points. The recording was made by projecting a beam of light on to the cornea whence it was reflected on to moving photographic film, and observations were carried out before and after performing a visual task for two hours. Some reduction in speed of movement after the work period was observed, but the amount of reduction did not correlate with increases in illumination from 5 through 100 to 300 lm/ft^2 (50 through 1000 to 3000 lux). The results were rejected as inconclusive.

More recently, Shackel (1967) has developed the oculographic technique for the measurement of eye movements. His technique has reached a high degree of sophistication and it is possible that, correctly applied, it offers a better opportunity than hitherto for the study of eye movements as a correlate of visual fatigue.

The blink rate, on the other hand, though of very early origin, has not been refined over the years, and although results of blink rate tests are often quoted, they are no longer considered to be useful objective measures of the factors which they set out to determine. The best known work in this field is probably that of Luckiesh (1944), who supported the validity of the test by the degree of consistency obtained in measurements in relation to such factors as the duration of the task, the size of the reading type, the presence of glare sources, etc. Luckiesh claimed that the great advantage of the method was that, being an involuntary function, performance cannot be improved deliberately by the subject trying to do better. Macpherson (1943) was, however, unable to repeat Luckiesh's results and, in fact, the disagreement was very much the same as that between Saldanha's fatigue investigation and that by the present authors (another typical case in which a technique which showed great promise and positive results in the hands of the original investigator failed to live up to this promise in the hands of others). Criticism of the blink rate test has been made that blink rate activity depends on many other factors apart from the state of physical or nervous fatigue, and Tinker (1947) has concluded that it is not a suitable criterion of ease of performance or of fatigue caused by continuous activity. Brozek and Simonson nevertheless included it in their investigation of visual fatigue, but found that there was difficulty in interpreting the results. Grandjean (1959) devised a visual motor reaction time test to assess the fatigue of telephone operators after a work shift. One of his tests consisted of reacting to a signal light by pressing a key, and although this did not measure a critical visual function, it may well be argued that the reaction time measured in this way could be expected to vary with the state of the central nervous system. Grandjean found that an increase in this 'optical reaction time' occurred for all of his fourteen subjects and was significant in all but one case.

Another test included by Grandjean was one of manual dexterity. This consisted of repeatedly dropping a plug through a tube as many times as possible in a period of two minutes. He claimed that a significant change was observed in the performance of this test by his telephone operators after their working shift.

Audiofrequency Threshold

It has been suggested that, in order to overcome the difficulty of resting the visual or central nervous system by changing the visual task to one capable of quantification, one should employ a test in another modality, such as hearing. One group of investigators (FORFA *et al.* 1956) used as a test the upper frequency threshold of audibility which had been found by Dirks and Horney (1954) to be related to fatigue. The measurement of this took about one minute and the test was carried out before work and in the course of every half-hour until the end of the shift, which occupied six hours with one twenty-minute break. The threshold frequency was found to fall steadily during working, with an intervening rise after the break, and the difference between the value at the beginning and the end of work was taken as a measure of the fatigue, induced during this period. The task given to the fifteen subjects in this investigation was the sorting and threading of beads on to wires. The illumination was provided by high-pressure mercury vapour lamps, by filament lamps, and by fluorescent lamps to levels of 3, 10, 30, 100 and 200 lm/ft^2 (33, 110, 330, 1100 and 2200 lux). There were slight differences between the results for the three different sources in that increase in illumination provided by the h.p.m.v. lamps did not increase work output as quickly as increased illumination provided by other sources, and that fatigue produced by fluorescent lamps was lower (for a given level of illumination) than that produced by the other sources. Taking the average result for the three sources, however, the usual form of increase of output with illumination level was shown, that is, the output rose with illumination level over the whole range, but the rate of increase fell off. Fatigue as defined by the drop in threshold audible frequency fell to a definite minimum in the region of 100 lm/ft^2 (about 1000 lux).

It is perhaps significant that this last finding is in complete agreement with observations of performance, performance decrement, and blink rate made by Brozek and Simonson referred to above.

Convergence Control

Abnormality in convergence muscle balance (heterophoria) has been looked for by Saldanha but he found no change in this function after his subjects had carried out the vernier setting test for two hours.

The relation between convergence control and fatigue may be one of cause rather than effect according to some measurements made by Michal (1954). Of those of a group of 154 office workers who complained of fatigue under fluorescent lighting, only 21% were found to have normal muscle balance, while of the other half of the group (who did not complain spontaneously of the lighting) 51% had normal muscle balance.

Accommodation Range

Michal (1954), following up ideas of Mercier (1952), also measured the available accommodation amplitude of his group of office workers and compared the average value for each age group after work under fluorescent lighting and after working by daylight. He claimed that the accommodation range was 'considerably lower' after working under fluorescent lighting, but his results are not detailed.

Baumgardt and Le Grand (1956) measured the near point of ten subjects before and after four hours of work under fluorescent lighting. In one case the fluorescent lamps were fed by single-phase alternating current, and in a second trial they were fed by direct current. On the average, a just significant reduction in the retreat of the near point was noted under direct current.

Accommodation Time

Following a discussion with Professor Le Grand on the above results and of the possible value of measurements on the accommodation mechanism as indicators of one of the possible manifestations of visual fatigue, the present authors devised a method of measuring accommodation time, and of the changes in accommodation time, which result during and after periods of visual work. Their apparatus (Collins and Pruen 1962, Fig. 6·4 and 6.5) consists of a means of viewing in rapid sequence a Landolt ring target apparently at or near infinite distance and one at or approaching the near point of distinct vision. This is achieved by seeing first, through a gap in a rotating sector disc, a Landolt ring printed in black on a clear film base at the focus of an achromatic lens and illuminated from behind, and secondly, via a mirrored portion of the sector, another ring closer to the eye. A shutter close to the eye gives a single exposure of the two targets, which are automatically changed before each exposure by stepping on by one frame the 35 mm film on which they are printed.

The subject was considered to have been capable of accommodating sufficiently fast if he correctly reported the orientation of the gaps in both the near and the far targets after more than five exposures out of ten at the same speed of the sector disc. The speed of the disc was gradually increased from a low value until this task could no longer be accomplished, and the shortest time for which the second target could be exposed was called the 'perception time'.

The subjects were first trained to use the apparatus and to make the judgments expected of them. When they had reached the limit of improvement in performance by practice, they were then put on to the experiment proper.

The visual task consisted of a two-hour session of setting the vernier gauge previously described above. Each subject removed their watches so that they had no knowledge of the time during the two-hour period they were seated at the vernier apparatus. The subject was instructed to set as many readings as he could as accurately as possible until he was told to stop. Perception times, that is, the time taken to change the accommodation from distant to near point, were measured before and after the two-hour period.

The results, though not wholly definite, demonstrated that perception time increased after visual work on twenty one occasions out of thirty nine, and decreased on eight occasions. On the remaining ten occasions no change was measured. Statistical analysis of variance suggested that there was no significant difference between the effects when the task was undertaken at a level of illumination of 1 lm/ft^2 (11 lux) and 30 lm/ft^2 (330 lux).

Fig. 6.4 *Accommodation-time measuring apparatus. Mounting of components*

Fig. 6.5 *Accommodation-time measuring apparatus, showing viewing position*

The results suggested that the effects of concentrated work over a period of two hours are likely to result in an increase in time taken to accommodate the eye and to recognise a small target. Although there were considerable differences between individual subjects, the average increase (regardless of illumination level) was statistically significant at the 5% level.

One of the most interesting aspects of the experiment was the subjective reactions. Most subjects remarked on the feeling of slight tiredness after a period of working on the vernier gauge. On twelve occasions this was identified as general tiredness, and on eight of these occasions there was also an increase in perception time. On five occasions subjects complained of both general and visual tiredness, and on two occasions they complained of visual tiredness only. Rather more complaints arose when the task was illuminated to 1 lm/ft^2 (11 lux) than when it was illuminated to 30 lm/ft^2 (330 lux).

This accommodation time or perception test has been used for the assessment of fatigue occasioned by an industrial task (coin inspection in the Royal Mint). The test does appear to give results which correlate with what would be expected from the nature of the tasks involved (Collins, Fox and Hopkinson 1964), but the correlation is by no means as positive and as close as is required for a fully quantitative study of industrial visual fatigue. The number of subjects whose results could be used in this study was reduced because several of the older workers were not able to achieve the initial amplitude of accommodation required by the apparatus, and refinements of the method to be introduced include an allowance to be made for the reduced accommodation distance of older subjects so that the accommodation amplitude can be related directly to each subject's capacity to accommodate.

Older people, in any case, take a longer time to accommodate. Robertson (1936) has shown an increase from 0·5 seconds for an age group of 20–24 to 0·7 seconds for an age group of 40–44, while Bannister and Pollock (1928–29) noted that one subject over 40 required an accommodation time of 0·62 seconds compared with two subjects in their early 20s who required 0·4 and 0·52 seconds respectively.

Since this work was carried out, a similar test has been developed in Japan by Matsui and Kondo (1963), and used to demonstrate the optimum illumination levels (between 50 and 200 lm/ft^2, or about 500 to 2000 lux) for various tasks. Krivohlavy, Kodat and Cizek (1969) have also measured speed of accommodation using two panels of Landolt rings at different distances.

Conclusion

Over a long period of activity by many research workers, of whom some have spent years of concentrated effort in the search for a fully valid and sensitive practical method of testing for visual fatigue or for fatigue visually occasioned, nothing has emerged which seems to be a substitute for direct subjective assessment of the situation. People will complain of unfavourable

visual conditions, and will take positive action against thoroughly unsuitable conditions, in circumstances under which the various quantitative objective tests barely register a positive, measurable effect. Of the various methods which have been discussed, it would appear that Grandjean's techniques of measuring the CFF have been developed to a stage of giving the most consistent results, but it must be borne in mind that, on at least two other occasions, satisfactory results by the originator of a technique have not been repeated by other investigators. The failure by the present authors to repeat Saldanha's results, and the failure by Macpherson and others to repeat the results on the blink rate test by Luckiesh and Moss, serve as warnings that until other investigators have demonstrated that Grandjean's techniques of CFF reduction measurements can be repeated consistently, no firm recommendation to adopt his methods can be put forward. It is also necessary to point out that in order to repeat Grandjean's study, highly refined techniques for the measurement of differences of flicker frequency of the order of only 1 or 2 cycles per second must be at the command of the research worker, quite apart from the need for techniques of proper statistical analysis of the data.

The particular phenomenon of flicker perception has received extensive study, and although some conflicting results have been obtained in relating it to various types of fatigue, the evidence on balance is in favour of flicker perception being related to the state of the central nervous system. There is evidently a need for work in this country to validate the relationship between psychophysical measurements of the kind discussed above and some agreed form of 'loading' of the nervous system.

It has been noted in Chapter 3 that many workers have concluded that an optimum illumination level for efficient and comfortable performance of most visual tasks occurs between 1000 and 2000 lux (100–200 lm/ft^2). This conclusion derives support from such measures of fatigue visually occasioned as have been successfully related to illumination level. These generally show a minimum of 'fatigue' in this region.

References

Bannister, H. and R. G. Pollock (1928–29): 'The Accommodation Time of the Eye.' *Brit. J. Psychol.* **19**, 394–396.

Bartlett, F. C. (1953): 'Psychological Criteria of Fatigue'. In 'Fatigue'. Ed. by W. F. Floyd and A. T. Welford. H. K. Lewis, London.

Baumgardt, E. and Y. Le Grand (1956): 'Sur La Fatigue Visuel.' Cahiers du Centre Scientifique et Technique du Batiment, No. 27, Cahier 230.

Bronner, A. (1967): 'L'Importance de l'Éclairage sur le Comportement Visuel de l'Enfant et de l'Adolescent.' *Lux*, **45**, 458–463.

Brozek, J. and A. Keys (1944): 'Flicker Fusion Frequency as a Test of Fatigue.' *J. Indust. Hygiene Toxicol*, **26**, 169–174.

Brozek, J. and E. L. Simonson (1952): 'Visual Performance and Fatigue under Conditions of Varied Illumination'. *Amer. J. Ophthalmol.*, **35**, 33–46.

Brozek, J., E. Simonson, and A. Keys (1950): 'Changes in Performance and in Ocular Functions Resulting from Strenuous Visual Inspection.' *Amer. J. Psychol.*, **63**, 51–66.

Carmichael, L. and W. F. Dearborn (1947): 'Reading and Visual Fatigue.' Houghton Mifflin, Boston.

Collins, J. B. (1959): 'Visual Fatigue and its Measurement.' *Ann. Occup. Hygiene*, **1**, 228–236.
Collins, J. B., J. G. Fox and R. G. Hopkinson (1964): 'Visual Fatigue.' *Ergonomics*, **7**, 363 (Abstract).
Collins, J. B. and B. Pruen (1962): 'Perception Time and Visual Fatigue.' *Ergonomics*, **5**, 533–538.
Dirks, H. and H. L. Horney (1954): 'Preliminary Information about a New Procedure for Determining Fatigue.' *Physiol. Rdsch.*, **5**, 317.
Ditchburn, R. W., D. H. Fender and Stella Mayne (1952): 'Vision with Controlled Movements of Retinal Image.' *J. Physiol.*, **145**, 98–107.
FORFA *et al.* (1956): 'Investigations into the Efficiency and Fatigue of the Human Being under Various Conditions of Illumination. (Untersuchungen über Leistung und Ermüdung des Menschen bei verschiedenen Lichtbedingungen).' *Lichttechnik*, **8**, 296–300. (Translation, BRS Library Communication No. 821, 1958.)
Grandjean, E. (1959): 'Physiologische Untersuchungen über die Nervöse Ermüdung bei Telephonistinnen und Büroangestellten.' *Internat. Z. Angew. Physiol. Arbeitsphysiol.*, **17**, 400–418.
Grandjean, E. and K. Bättig (1955): 'Das Verhalten des subjectiven Verschmelzungsschwellen des Auges unter Verschiedenen Arbeits- und Versuchsbedingungen'. *Helv. Physiol. Pharmacol. Acta*, **13**, 178–190. (Translation, BRS Library Communication No. 766.)
Grandjean, E., R. Egli, F. Diday, W. Block and H. Gfeller (1953): 'Die Verschmelzungsfrequenz intermittierender Lichtreize als Ermüdungsmass.' *Helv. Physiol. Pharmacol. Acta*, **11**, 355–360.
Grandjean, E. and H. W. Jaun (1960): 'Ermüdungsmessungen bei Telephonistinnen während der Nachtarbeit.' *Z. Präventivmed.*, **5**, 143–152.
Grandjean, E. and E. Perret (1961): 'Effects of Pupil Aperture and Time of Exposure on the Fatigue-Induced Variations of the FFF'. *Ergonomics*, **4**, 17–23.
Khek, J. and J. Krivohlavy (1968): 'Evaluation of the Criteria to Measure the Suitability of Visual Conditions.' Proc. CIE 16th Session (Washington, 1967), Paper P, 67–19. Commission Internationale de l'Éclairage, Paris.
Krivohlavy, J., V. Kodat and P. Cizek (1969): 'Visual Efficiency and Fatigue During the Afternoon Shift.' *Ergonomics*. (To be published.)
Lancaster, W. B. (1932): 'Ocular Symptoms of Faulty Illumination.' *Amer. J. Ophthalmol.*, **15**, 783.
Le Grand, Y. (1961): 'Effets Physiologiques de la Couleur—Application Spéciale au Cas des Écoles.' Cahiers du Centre Scientifique et Technique du Bâtiment, No. 50, Cahier 404, 7–10.
Lion, K. S. (1952): 'Oculometric Muscle Forces and Fatigue.' *Illum. Engng.* (*New York*), **47**, 388–390.
Lioublina, E. I. (1944). 'Problemes de l'Optique Physiologique.' Edit. Acad. Sci. URSS, **2**, 198–203.
Lord, M. P. and W. D. Wright (1950): 'The Investigation of Eye Movements.' *Rep. Progr. Phys.*, **13**, 1–23.
Löwenstein, O. and I. Löwenfeld (1951): *Arch. Neurolog.*, **66**, 580–599.
Löwenstein, O. and I. Löwenfeld (1952): *J. Nervous Mental Disease*, **115**, 1–21 aod 121–145.
Luckiesh, M. (1944): 'Light, Vision and Seeing.' pp. 205–212. Van Nostrand, New York.
Luckiesh, M. and F. K. Moss (1933): 'A Correlation Between Illumination Intensity and Nervous Muscular Tension Resulting from Visual Effort.' *J. Exper. Psychol.*, **16**, 540.
Macpherson, S. J. (1943): 'The Effectiveness of Lighting: Its Numerical Assessment. 3. By Methods Based on Blinking Rates.' *Trans. Illum. Engng. Soc.* (*London*), **8**, 48–49.
Matsui, M. and M. Konda (1963): 'Studies on Relation Between Illumination Levels and Visual Fatigue.' *J. Illum. Engng. Inst. Japan*, **47**, 176–186. (In Japanese with English Summary.)
Mercier, A. (1952): 'L'Éclairage Fluorescent, Est-il Nocif pour la Vision?' *Ann. d'Oculistique*, **185**, 577–587.
Michal, F. V. (1954): 'Visual Fatigue.' *Csl. Opthalmol.*, **10**, 362–367. (Translation, BRS Library Communication No. 795, 1957.)
Motokawa, K. and K. Suzuki (1948): 'A New Method for Measuring Fatigue.' *Jap. Med. J.*, **1**, 200–206.
Robertson, C. J. (1936). 'Measurement of Speed of Adjustment of the Eye to Near and Far Vision.' *Arch. Ophthalmol.*, **15**, 423–434.
Ryan, T. H., M. E. Bitterman and C. L. Cottrell (1953): 'Relation of Critical Fusion Frequency to Fatigue in Reading.' *Illum. Engng.* (*New York*), **68**, 385–391.
Saldanha, E. L. (1955): 'An Investigation into the Effects of Prolonged and Exacting Visual Work.' Medical Research Council Report, APU 243/55, October.
Schneider, L. (1963): 'Optimale Beleuchtungsstärken fur die Arbeit.' *Lichttechnik*, **15**, 617.

Ségal, J. (1950): 'Les Effets de l'Éclairage par Tubes à Fluorescence sur la Fatigue Visuelle.' Cahiers du Centre Scientifique et Technique du Bâtiment, Cahier 84, 1–8.
Shackel, B. (1967): 'Eye Movement Recording by Electro-Oculography.' Chap. 9 in 'A Manual of Psychophysical Methods.' Ed. by P. H. Venables and I. Martin. North-Holland Publishing Co., Amsterdam.
Simonson, E. and J. Brozek (1948A): 'Effects of Illumination Level on Visual Performance and Fatigue.' *J. Opt. Soc. Amer.*, **38**, 384–397.
Simonson, E. and J. Brozek (1948B): 'The Effect of Spectral Quality of Light on Visual Performance and Fatigue.' *J. Opt. Soc. Amer.*, **38**, 830–840.
Simonson, E. and J. Brozek (1952): 'Work, Vision, and Illumination.' *Illum. Engng.* (*New York*), **47**, 335–349.
Simonson, E. and N. Enzer (1941): 'Measurement of Fusion Frequency of Flicker as a Test for Fatigue of the Central Nervous System.' *J. Indust. Hygiene Toxicol.*, **23**, 83–89.
Snell, P. A. (1933): 'An Introduction to the Experimental Study of Visual Fatigue.' *J. Soc. Mot. Pict. Engnrs.*, May, 367–390.
Tinker, M. A. (1947): 'Illumination Standards for Effective and Easy Seeing.' *Psychol. Bull.*, **44**, 435–450.
Werner, M. and E. Grandjean (1960): 'Beeinflussung physiologischer Vorgänge und psychischer Leistungen durch Barbitalum solubile (Veronal).' *Helv. Physiol. Pharmacol. Acta*, **18**, 225–240.
Weston, H. C. (1945): 'The Relation between Illumination and Visual Performance.' Industrial Health Research Board Report No. 87 (reprinted 1953). HMSO, London.
Weston, H. C. (1953): 'Visual Fatigue with Special Reference to Lighting.' *Trans. Illum. Engng. Soc.* (*London*), **18**, 39–66.
Weston, H. C. (1962): 'Sight, Light and Work.' H. K. Lewis, London.

7

Light Source Colour and Colour Rendering

For a clear understanding of the problems involved in the prediction of the rendering of colours by various sources of light, or for the specification of sources to achieve a desired appearance of colours, it is necessary to invoke the basic principles of colour vision and colorimetry. These are set out in detail elsewhere (e.g. Wright 1964). For the purpose of the present discussion, the underlying principles of colour rendering are set out below.

Current ideas on colour measurement and colour rendering are based on the trichromatic theory of colour vision known as the Young–Helmholtz theory. This theory states that any colour perceived by the eye can be matched by a mixture of appropriate amounts of light from three primary colours. It can be demonstrated experimentally that most of the colours of nature can be matched by mixtures of red, green and blue as primaries, and more recently it has been demonstrated that there appear to be three separate and distinct types of receptor in the eye which appear to correspond in their spectral sensitivity to these three primaries. There is therefore at the moment a consensus of opinion (though not unanimous) that the Young–Helmholtz trichromatic theory of colour vision is fully validated. Certainly the majority of colour phenomena met with in lighting practice can be satisfactorily handled in quantitative terms by the assumption of a trichromatic basis for the numerical exercise.

Based on this three-mechanism process, the Commission Internationale de l'Eclairage (CIE) has standardised on a method of colour specification by which any colour is represented by its position on a tricoordinate system, the corners of this 'colour triangle' representing the positions of the three CIE primary colours. These CIE primaries are, in fact, theoretical, because it has been found experimentally that all possible recognisable colours cannot, in fact, be obtained by mixtures of any real primaries, so to this extent the Young–Helmholtz theory has been proved inadequate. Nevertheless, by the assumption of three theoretical primaries (corresponding approximately to red, green and blue sensations) chosen so that all possible attainable colours lie inside the equilateral triangle joining these three theoretical primaries on

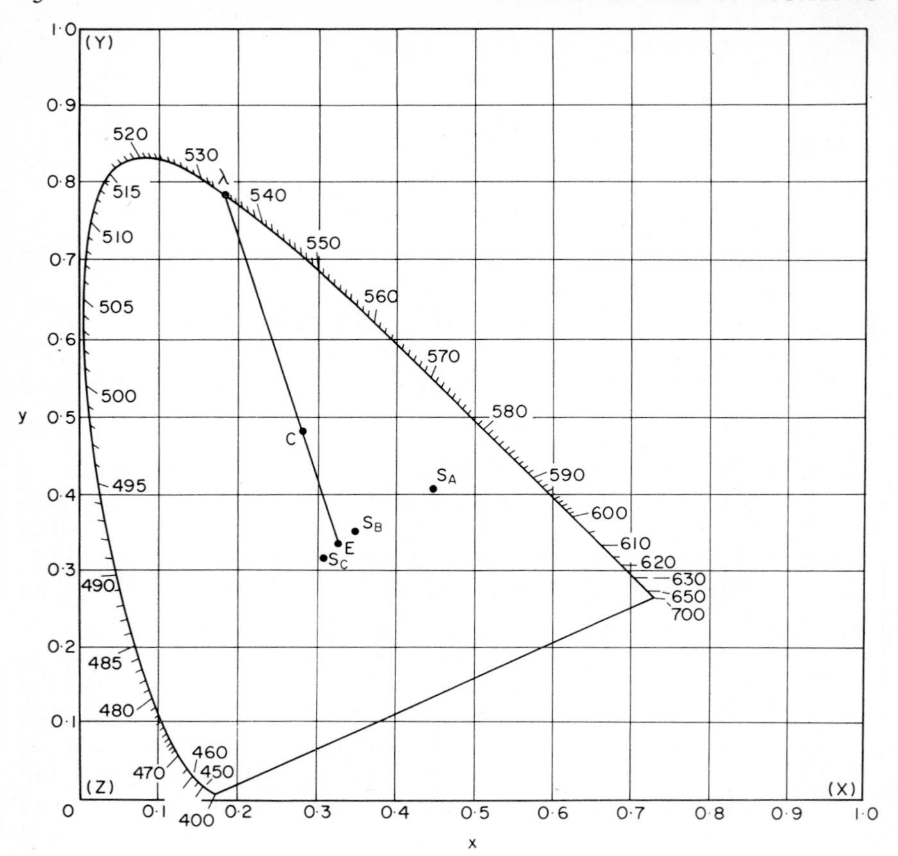

Fig. 7.1 *The 1931 CIE chromaticity chart plotted in terms of the reference stimuli (X), (Y), and (Z) with units based on an equal-energy white E.*
The location of the CIE standard illuminants S_A, S_B and S_C is shown (From Wright 1964)

the CIE coordinate scale, a consistent system of colour specification and colorimetry can be devised.

The CIE primaries are known as X, Y, and Z, and the coordinates of any points inside the triangle are designated in terms of x, y and z. Thus the centroid of the triangle, representing equal mixtures of all three primaries, and therefore giving an achromatic sensation (white, grey), is given by $z = 0{\cdot}333$, $y = 0{\cdot}333$, $z = 0{\cdot}333$. Since the coordinate system is chosen so that $x + y + z$ always adds up to unity, it is consequently only necessary to specify two of the three parameters, usually x and y, so that the diagram can be plotted on rectangular coordinates as shown in Fig. 7.1, and the locus of fully saturated spectrum colours is as shown by the inverted U-shaped curve. In this diagram E is the equal energy white point (i.e. the neutral achromatic point) and a colour represented by point C will have a hue represented by the intersection of the extension of the line EC on the spectrum locus. The saturation (i.e. the intensity or colourfulness) of the colour is given by the relationship of the distance YC to the distance CE. In other words, the nearer C is to E, the less saturated (that is, the paler or less colourful) the colour will be, while if C is near to the spectrum locus, it will be highly saturated, that is, a purer or more intense colour. The spectrum locus is the limiting boundary of all possible colour sensations, and is closed by a line joining its two extremities. It is an interesting and useful property of the eye that light of quite a large range of chromaticity (see below) around the neutral point is interpreted visually as 'white', and that this range can be extended by allowing the eye time to become adapted to the light in question. This is the reason why, for example, natural daylight in all its various manifestations (blue sky, overcast sky, direct sunlight etc.) and various forms of 'white' artificial light (light from incandescent filament, light from various forms of fluorescent lamp) appear to the eye when fully adapted as 'white' though their coordinates on the CIE system differ widely.

In considering the colour attributes of the light source, two properties must be discussed. These are (a) chromaticity, and (b) spectral distribution of the radiated power. The two are not linked unequivocally, because although the spectral power distribution of a source will determine its chromaticity on the CIE diagram, the reverse is not true.

The first man-made light sources were obtained by burning organic materials (wood, candles, oil lamps), and these sources, together with the modern incandescent filament lamp, all have the common property that they radiate light continuously throughout the whole of the visible spectrum. In addition, their spectral power distribution corresponds very closely to the theoretically perfect radiator (the 'black-body' radiator), that is, one which is luminous only by virtue of the radiation which it emits, rather than that which it reflects. This black-body radiator is sometimes called a 'full radiator' or a 'Planckian' radiator. Planckian or black-body sources do not radiate energy equally throughout the spectrum, however; the curve showing relative energy at each point in the spectrum will vary in shape as the temperature of the radiator varies (Fig. 7.2). At low temperatures the majority of the energy will be radiated in the infrared region of the spectrum, and so little or no visible emission of light will occur. At higher temperatures, however, progressively more visible radiation will result, at first a dull red glow, then with increasing temperature, say 2000 K (Kelvin or Absolute Temperature), the light will be a reddish white, while at 3000 K, the light will appear to the adapted eye as fully white, although the maximum emission of energy will, in fact, still be in the infrared region. A tungsten filament in a gas-filled general service lamp running at this temperature will not give quite the same coloured light owing to the fact that its emissivity is not quite the same as that of a theoretical black body at the same temperature. The actual temperature of the lamp filament can, however, be adjusted so as to make the colour of the light emitted by the lamp an exact match with the black-body radiator at a particular temperature. This black-body temperature is then

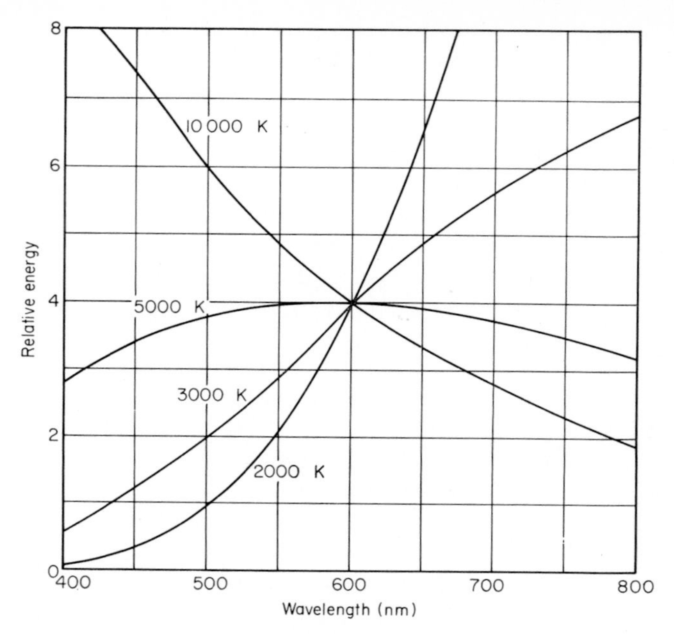

Fig. 7.2 *Spectral distribution of energy emitted by Planckian (black-body) radiators at various temperatures (From Wright 1964)*

called the 'correlated colour temperature' of the light emitted by the filament lamp.

At a temperature around 3650 K, tungsten melts and this sets a limit to the spectral power distribution characteristic of an incandescent filament lamp.

The spectral distribution of radiated power of a black-body radiator at any temperature can be calculated from Planck's Law, and so it can be demonstrated that the peak radiation for a black-body radiator is around 600 nanometres (nm), that is, in the orange region of the spectrum, when the colour temperature is 5000 K, and in the ultraviolet region (beyond 400 nm) when the colour temperature is 10 000 K. Sources of higher colour temperature than that attainable by tungsten filament lamps result from other means of generating light, such as the various forms of discharge lamp with or without the aid of fluorescence. In addition, the use of appropriate transmitting colour filters, which suppress the longer wavelength end of the spectrum in varying amounts, permits light of varying high correlated colour temperature to be obtained from an incandescent source. Various forms of natural daylight have very high correlated colour temperatures. Measurements made by Collins (1965) and by Henderson and Hodgkiss (1963, 1964) give the frequency of occurrence of various chromaticity values for natural daylight related to the correlated colour temperature. The correlated colour temperature of daylight varies considerably from time to time according to the weather and the amount of sunlight, direct or reflected, from earth and cloud. This inconstancy of the colour of daylight has great practical importance and will be discussed later.

Since the change in colour temperature of a full radiator results in a smooth transition of colour from red through orange, yellow and neutral white to a bluish white, this change can be represented on the CIE chromaticity diagram as a curve of the form shown in Fig. 7.3. This curve is called the Black-Body or Planckian Locus, and serves as a useful frame of reference when discussing the appearance of near-white light sources. If the source is approximately a black-body radiator, its position in the chromaticity diagram

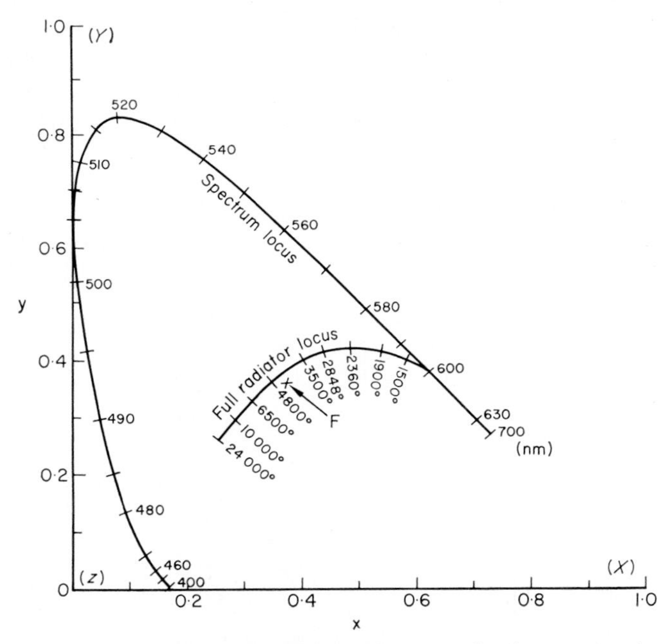

Fig. 7.3 *The locus of Planckian (black-body) radiators in the chromaticity chart. Absolute temperatures are shown against the locus (From Wright 1964)*

and also its spectral power distribution will be defined adequately merely by specifying its correlated colour temperature. If, on the other hand, the chromaticity of the source is such that it falls just to one side of the Planckian locus, we can tell that it will appear similar in colour to a black-body at the temperature corresponding to the point of the Planckian locus nearest to the point representing its chromaticity. Kelly (1963), following work by Judd, has shown how to locate this point in relation to a series of 'isotemperature' lines, given the chromaticity coordinates of the source in question. This point gives the value of the correlated colour temperature for the source with these coordinates. If the chromaticity is known to be above the Planckian locus, it can be predicted that the source will appear somewhat greener than the corresponding black-body radiator of the same colour temperature, while if it is below the Planckian locus, the source will appear pinker.

From what has been said earlier about the equivalence of any one colour to mixtures of three primaries, it will be appreciated that a white or near-white light of any colour temperature does not have to be produced by a source with a continuous spectrum of radiation corresponding to that of the Planckian radiator, since it could be produced by an appropriate mixture of, for example, single spectrum colours. Consequently it is not possible to predict the shape of the spectral power distribution curve merely by knowing the correlated temperature, unless, of course, we know that the characteristics of the source are such that it corresponds closely to a black-body radiator. As an example, the chromaticity of the light from a fluorescent lamp of the type known as 'daylight' may fall at a point F on the CIE chromaticity diagram as shown in Fig. 7.3, but, in fact, the spectral power distribution, as shown in Fig. 7.4, is far from that of a Planckian black-body radiator of the same correlated colour temperature.

The chromaticity of a light source can only determine the appearance of the source itself, or of a neutral (achromatic) coloured surface illuminated by the source. In order to predict the colour sensation resulting from coloured reflecting surfaces as seen under different light sources, it is necessary to con-

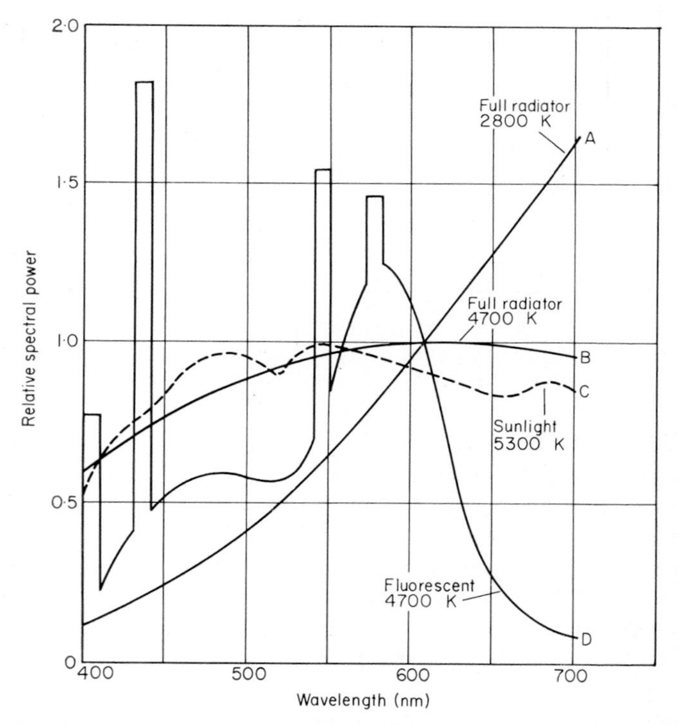

Fig. 7.4 *Distribution of power over the visible spectrum for four types of light source (Adapted from Crawford 1963)*

sider the spectral power distribution of the light source in conjunction with the spectral reflecting characteristics of the surface.

Since an object cannot reflect *more* light of any wavelength than falls upon it, it can only appear coloured by virtue of its absorption of light of colours other than that which it reflects. Therefore, if there is no light in the illuminant of a colour which the object reflects, the surface must necessarily appear dark. A red London bus or a green country bus each appears dark grey when illuminated by monochromatic yellow sodium discharge lighting radiation because the low-pressure sodium lamp emits at only one wavelength, which is indifferently reflected as the same wavelength by both the red paint and the green paint of both buses.

Again, what is subjectively a white light need not necessarily have a continuous spectral power distribution in order to appear white. (White light can arise from the correct mixture of, for example, red light with a wavelength of 700 nm and the complementary peacock-blue light of approximately 495 nm.) A white surface lit by this white source will look white. If this source is used to illuminate a yellow object, however, which absorbs strongly in the blue region of the spectrum, but which allows the reflection of yellow, red and green light, the yellow object will appear reddish because there will be a strong reflection of the red light component of the illuminating source, but the peacock-blue component will not be reflected. The object will therefore appear to be quite a different colour from that seen under light of a black-body radiator of identical chromaticity (and therefore of identical subjective colour appearance). This phenomenon is known as 'metamerism'.

Colour Adaptation

In theory, the colour of any object will change with the colour, or colour temperature, of the light source which illuminates it. There is, however, an important visual phenomenon, of great practical importance, which modifies such changes and limits their noticeability. This phenomenon is known as *colour adaptation* and is one of the factors (the other is known as *colour constancy*) which ensures that light of chromaticity over a large area of the CIE chromaticity diagram will still appear white after the eyes are completely adapted to it. This time for adaptation is important. If one near-white source is compared with another directly, either juxtaposed in space or presented very quickly one after another before colour adaptation can take place, very small differences in chromaticity can be detected visually.

Colour constancy is the name given to the psychological mechanism which makes us see a colour as we think it ought to be rather than as it is. For example, the white ceiling of a room with a sunlit lawn outside the window will be quite green in the area near the window, but it looks to be the same white colour all over, due to the colour constancy effect.

Objects illuminated by light sources with a smooth and continuous spectral distribution, as with a black-body radiator, will appear very much the same under a wide range of colour temperatures. The majority of ordinary everyday objects have very much the same colour appearance under light sources as widely different as a tungsten filament lamp and average natural daylight. Close observation is necessary to detect the difference. It is only with objects of specially sensitive colour properties, such as some foods or textiles, that the difference becomes troublesome.

Complete colour adaptation of this kind is only possible between sources with a smooth continuous spectral power distribution of the black-body or near black-body type of radiator. Most commercial fluorescent lamps, on the other hand, have spectral power distributions appreciably different from the black-body type, and, as will be seen from Fig. 7.4, the characteristic fluorescent lamp spectral power distribution consists of a number of bands in which the mercury discharge emits its energy, joined by a continuum of light emitted by the fluorescent coating. As might be expected, the design of a fluorescent lamp to give apparently white light of any appearance from 'cold' to 'warm' (i.e. bluish or reddish in character) is a relatively simple matter. There are a great many possible combinations of fluorescent materials which permit light from various parts of the spectrum to be added to the basic mercury vapour discharge emission. Since, however, chromaticity (governing colour appearance) and spectral power distribution are not unequivocally related, it does not follow that the achievement of the desired colour appearance in a fluorescent lamp will also result in a spectral power distribution of the desirable emission characteristics to relate satisfactorily to the spectral reflection characteristics of the materials and objects to be illuminated in a lighted interior. In order to achieve a smooth Planckian spectral power distribution curve, it would be necessary to suppress the characteristic mercury discharge bands, or alternatively to swamp these bands by the relative increase of the output of the fluorescent material. The only lamps which have approached the ideal spectral power distribution curve use special techniques, which may involve the use of a special double coating of luminescent material, sometimes in combination with a layer of absorbent material to filter out the unwanted mercury discharge radiation. At the moment these expedients result in a considerable reduction in overall light output of the lamp, so that it becomes relatively inefficient as compared with a normal commercial type of lamp. For the majority of interior lighting purposes, the colour rendering of objects may be relatively unimportant, and so almost the whole manufacturing output of fluorescent lamps is confined to the lamps of the more efficient type, reserving the special lamps with a near-Planckian spectral power distribution curve for special purposes where a high accuracy of colour rendering is essential.

Colour Appearance and Colour Rendering

The *colour appearance* of a light source is therefore defined by the chromaticity coordinates on the CIE diagram of the light emitted by the source, and this same chromaticity applies to the appearance of a white or neutral surface which receives and reflects light from the source. The colour rendering given by the source, on the other hand, depends upon its spectral power distribution. Two light sources of the same chromaticity, one from a Planckian radiator and the other from a discontinuous radiator like a discharge lamp, will appear of the same colour themselves, and will render the colour of a white or neutral reflecting surface exactly the same, but their effect upon any coloured object or material can, and probably will, be entirely different. The colour appearance of a light source is therefore defined by its chromaticity, but the colour rendering which it will give depends upon its own spectral power distribution and upon the spectral reflecting characteristics of the material or objects which it is illuminating.

The colour rendering properties of a light source can usefully be considered under two headings. First, there is the 'accuracy' of the rendering given by- the light source as compared with a familiar light source such as average daylight. In addition to this 'true' colour rendering, there is the 'preferred' colour rendering, for example, its warmth or cheerfulness as compared with the 'cold light of day'. The preferred colour rendering appears to be related to the level of illumination. Kruithof (1941) confirmed experimentally the well-known phenomenon that at low levels of illumination, most people prefer a 'warm' light, whereas at high levels of illumination, a 'cold' light can be tolerated and may even be preferable. It is widely believed that this is because we are used to high levels of illumination from cold, natural daylight, whereas in the interior of our homes long experience with warm light from the fire or from incandescent lighting such as candle light or incandescent filament light, is related in our minds with low levels of illumination. A psychological explanation of this kind may, in fact, be sufficient, but further experiment may reveal a physiological basis for this form of colour preference. It should be noted, however, that it appears to be well attested that in tropical regions, the indigenous population prefers fluorescent lighting of a different (higher) correlated colour temperature from the domiciled Europeans. This again may be background culture, or it may be due to the different retinal pigmentation of the pale and dark skinned peoples.

Natural daylight is generally preferred for various inspection processes of textiles, foodstuffs, colour printing materials, pigments etc. Unfortunately natural daylight is a variable source in terms of colour temperature, spectral power distribution, and, of course, the level of illumination. When highly sensitive colour judgments have to be made, it is difficult or impossible to ensure that the natural daylight under which colour judgments are made is sufficiently constant to prevent error.

For this reason the British Standards Institution has developed a standard for an artificial source which corresponds sufficiently closely to daylight for colour matching and colour rendering processes. In a recent revision of this standard (BS 950:1957), an attempt was first made to arrive at an agreed value for the most commonly used colour temperature of daylight. It was found that a wide range of values covering the range from 5000 K to 7500 K was commonly claimed to be used, and in considering the data of frequency of occurrence obtained by Collins (1965) it was decided to recommend standardisation on a colour temperature of 6500 K, which is close to the mean value of the correlated colour temperature of the average light from the whole sky (rather than pure north sky light but still excluding sunlight).

The spectral power distribution for this standard colour of daylight was obtained from extensive measurements by Henderson and Hodgkiss (1963, 1964), and Condit and Grum (1964) who demonstrated that there is, in fact, a fairly constant relationship between the spectral distribution and the colour temperature of natural daylight over a very wide range of apparent colour.

For the lighting of interiors at night, a warmer colour of light than that provided by daylight is universally preferred, particularly in places intended for relaxation or social activity. This preference is certainly related, so far as this country and others with similar cultures are concerned, to the long tradition which shows itself in the pleasure of dining by candlelight or talking to friends by the light of an open fire.

Fig. 7.5 shows the relationship obtained by Kruithof, referred to above,

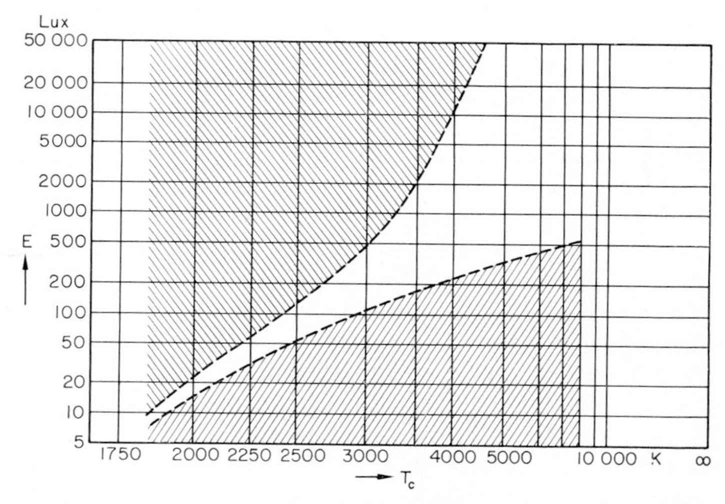

Fig. 7.5 *Relation between colour temperature and illumination* (*From Kruithof 1941*)

For every colour temperature there exists a highest and lowest level of illumination at which the illumination is considered 'pleasing': at lower levels the illumination appears dim or cold, at higher levels the colour rendering is unnatural. The left-hand part of the limiting curves, up to a colour temperature of 2850 K, is recorded by allowing electric lamps with variable (decreased) current to burn in a room, and varying the number of lamps. The illumination intensity on a table 80 cm high was here measured. In the right-hand part the lowest level which does not give the impression of coldness was determined by experiments with daylight itself and with daylight luminescence lamps. The shape of the upper curve has been extrapolated in this region with the help of the fact that in direct sunlight (colour temperature 5000 K), even with the highest illumination intensities occurring (10^4 or 10^5 lux), the colour rendering is never found 'unnatural'. On the abscissa the reciprocal value of the colour temperature Tc is plotted, on the ordinate the logarithm of the illumination intensity E, since $1/Tc$ and $\log E$ are measures of the physiological estimation of these quantities. In these coordinates the lower limiting curve takes on a nearly linear form. It may be mentioned that the experiments were carried out in a laboratory room. It was found, however, that in a living room with light-coloured furniture and wall coverings roughly the same limits are obtained.

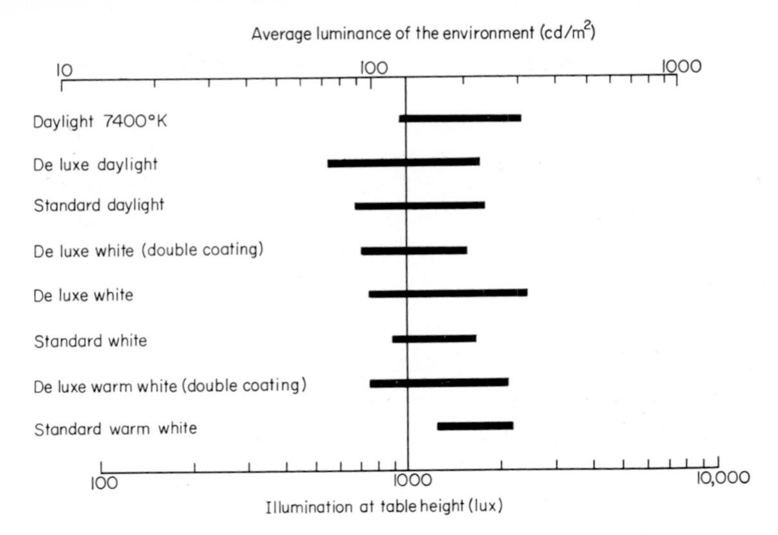

Fig. 7.6 *Subjective evaluation of satisfactory luminance levels in a conference room illuminated by different fluorescent lamps.* (*From Bodmann 1967*)

(A maximum percentage of 'good' ratings—based on a probability rate of 90%—occur within the indicated ranges)

Impressions associated with different levels and colours of fluorescent lighting in a conference room

Average level		*Colour of light*		
Illumination (lux)	*Luminance (cd/m²)*	*Warm white*	*White*	*Daylight*
Less than 700	90	Not unpleasant	Dim	Cool
700–3000	90–380	Pleasant	Pleasant	Neutral
More than 3000	380	Excessive, artificial	Pleasant, lively	Pleasant

between illumination level and colour temperature of the light source. It will be seen that according to these studies, it would be expected that a light of colour temperature 6500 K would have to be related to an illumination level of not less than 500 lux, while low illumination levels of the order of 100 lux are only pleasant if provided by light of colour temperature not greater than 3000 K. This relationship has been further studied by Bodmann and his colleagues (1964), and the tendency has been to some extent confirmed over a range of 600 lux (56 lm/ft^2) to 3000 lux (280 lm/ft^2) (see Fig. 7.6).

It was probably due to this phenomenon that fluorescent lighting obtained a bad reputation in its early days. The first commercial lamps to be widely available (known as 'daylight' type lamps) suffered from such a marked lack of red light, and consequently such a cold appearance, that the correlated colour temperature was excessively high for the levels of illumination which could then be provided under the prevalent economic conditions. Subsequent developments have led to major improvements. Lamp research by the manufacturers has enabled a warmer colour of lamp to be provided, where necessary, for normal interior lighting, so that lighting more in line with Kruithof's findings can be designed. In addition, very much higher levels of illumination can be and are afforded so that a proper use can be found for lamps with a high correlated colour temperature, particularly for use as a supplement to natural daylight (permanent supplementary artificial lighting—see Chapter 11). Lamp research has also been directed to producing lamps of special colour rendering properties, such as the artificial daylight standard referred to above for use where accurate colour rendering and colour matching is necessary.

Colour Rendering and Luminous Efficiency

The 'efficiency' (or to use the more recently standardised term 'efficacy') of any light source is expressed in lumens per watt, where the lumen as a unit of visible radiation at different wavelengths is defined in relation to the curve of relative spectral luminosity of the radiation. The CIE has standardised this relative spectral luminosity function, in terms of an international agreement on the spectral characteristics of human vision, as determined from extensive psychophysical experiments. This CIE standard luminosity function is shown in Fig. 2.1. This shows that the maximum spectral luminosity occurs at 555 nm in the yellow-green region, and so on the basis of the definition of lamp efficacy, a lamp radiating its whole output at or around this wavelength (for example, a low-pressure sodium discharge lamp) must inevitably have a higher luminous efficacy than a lamp which radiates the same amount of visible light spread evenly throughout the spectrum. Basically, therefore, any attempt to develop a fluorescent lamp to radiate throughout the spectrum and match the spectral power distribution curve of a Planckian radiator, and so give widely acceptable colour rendering properties, will be penalised so far as luminous efficacy is concerned as compared with the lamp designed to emit most of its light in the region where the eye is most sensitive.

It is clearly open to question whether the present method of defining the luminous efficacy of a light source, without reference to the performance of the light source in important aspects of practical vision, should not be modified in some way. The present definition of luminous efficacy is only justified in the most elementary psychophysical terms. It serves as a correlate of the performance of the eye on a black and white or neutral visual task, but for all other practical situations it is irrelevant and penalises the good spectral power distribution to the advantage of the bad.

Developments in colour rendering theory and practice are closely tied up with the commercial development of the fluorescent lamp. Following the original unsatisfactory 'daylight' lamp, the first new introduction was a 'warm white' fluorescent lamp in which the luminous efficacy of the lamp (as currently defined) was maintained by means of high output in the yellow region of the spectrum, but by virtue of a greater amount of light towards the red and less in the blue region of the spectrum, the overall colour appearance approached more closely that of an incandescent filament lamp. The warm white lamp was therefore recommended for use in interiors after dark in preference to the daylight type lamp, and it was certainly an improvement. Nevertheless the colour rendering of this new lamp was still not satisfactory, and further attempts were made to improve the situation by introducing more radiation at the red end of the spectrum, where it was still seriously lacking. These types of lamp with added red emission (and consequently lower luminous efficacy) were all designated by the term 'de luxe', and they were introduced to satisfy those who wanted a reasonable standard of colour rendering and were prepared to sacrifice something in luminous efficacy.

Various manufacturers adopted different names for their ranges of lamps, covering those of high luminous efficacy and poor colour rendering, and of low efficacy with improved colour rendering. Eventually the British Standards Institution introduced a standard (latest revision—BS 1853:1967) which attempted some control over the performance and colour rendering of the confused variety of lamps on the market. The colour appearance of five classes of fluorescent lamp is now specified by this British standard in terms of CIE chromaticity coordinates with an appropriate tolerance in the form of an area surrounding the desired point, but unfortunately these tolerances cannot be made so small that differences between lamps within the specification cannot be detected visually.

This situation is inherent in present difficulties of manufacture. It accounts for the fact that replacement lamps, even of the same type by the same manufacturer, can be seen to be different. Eventually, with improved manufacturing techniques, the situation is likely to improve. As a result of this British standard, the designations commonly employed by manufacturers, for example, 'daylight', warm white', etc., now give some clear indication of the chromaticity of the light. See Table 7.1.

The British Standard also attempts to specify the colour rendering properties of two of the types of lamps. The method employed is an abridgment of the spectral power distribution of the light emitted by the lamp. The visible spectrum is divided into eight bands, and the target and appropriate tolerances on the amount of light to be emitted in each of these eight bands is specified (see Table 7.1).

Colour Rendering of Surfaces

Theoretically it is possible to predict the change of appearance of a coloured surface under two light sources having different spectral power emission distributions, provided the spectral reflection characteristics of the surface are known, and if allowance is also made for the effects of colour adaptation. This theoretical prediction method is, in fact, the basis of a proposal (originally made by Judd and his associates) for an international method of specifica-

Table 7.1

COLOUR, COLOUR RENDERING, AND LIGHT OUTPUT FROM FLUORESCENT (MCF/U) LAMPS. FROM BRITISH STANDARD 1853:1967

				Colour Designation				
				Colour Matching or North Light	*Daylight*	*Natural*	*White*	*Warm White*
Nominal chromaticity coordinates	High Loading Lamps		*x*	0·313	0·365	0·375	0·405	0·436
	125 W, 80 W, 30 W (1 in dia.)		*y*	0·328	0·373	0·360	0·391	0·404
	Low Loading Lamps		*x*	0·313	0·368	0·379	0·410	0·442
	20 W, 40 W, 30 W (1½ in dia.)		*y*	0·323	0·371	0·358	0·389	0·402
	CIE band wavelength limits (nm)							
Percentage luminance in spectral bands	1.	Far Violet	380–420	0·017 min.		0·01 min.		
	2.	Violet	420–440	0·42 min.		0·30 min.		
	3.	Blue	440–460	0·56 min.		0·30 min.		
	4.	Blue-Green	460–510	8·1 min.		5·0 min.		
	5.	Green	510–560	45·8 max.		39·0 max.		
	6.	Yellow	560–610	39·0 max.		51·0 max.		
	7.	Light Red	610–660	8·0 min.		8·5 min.		
	8.	Dark Red	660–760	0·22 min.		0·2 min.		
	Length (in)	*Diam. (in)*	*Rated power (W)*					
*Lighting design lumens**	24	1½	20	650	1000	750	1050	1050
	36	1	30	1150	1700	1300	1750	1750
	36	1½	30	1100	1650	1300	1700	1700
	48	1½	40	1700	2500	1950	2600	2700
	60	1½	65	2700	4200	3100	4400	4400
	60	1½	80	3100	4650	3600	4850	4850
	96	1½	125	5150	7900	5950	8300	8300

* Luminous flux at 2000 hours, approximating to average lumens throughout 5000 hours of life.

tion of colour rendering of a source, by which the changes in chromaticity of a number of selected surface colours are calculated for the source with reference to a standard.

While an objective method of this type (which we shall discuss later) is of value, provided that the meaning of the chromaticity shifts can be usefully interpreted, it has a drawback in that the calculations are highly complex and must be processed by an electronic computer. There is, however, an alternative and more direct method which is based on subjective assessments of acceptable change in colour rendering. A pilot study of this kind was conducted at the Building Research Station (Gloag 1961) in which five different fluorescent lamps were studied, and the change in appearance of each of the colours in the British Standard colour range (BS 2660:1961) was recorded when each of the five lamps in turn were substituted for natural daylight. A three-point scale of assessment was used (no perceptible change, perceptible change but acceptable, and unacceptable change), and from the results it was possible to predict which of the five lamps would be most generally acceptable for its colour rendering characteristics, as well as indicating which colours would have to be avoided when any one of the five lamps was used. Broadly, the results showed that the 'colour-matching' type of lamp was most generally acceptable, closely followed by two improved forms of 'daylight' lamp, with 'warm white' and 'warm white de luxe' placed last. It must be remembered, however, that 'warm white de luxe' is deliberately designed to give a colour temperature approaching that of a tungsten filament lamp, which is itself a long way removed from daylight. If the experiment had used incandescent filament lighting as the standard rather than natural daylight, the order would certainly have been changed and possibly reversed. This emphasises that colour rendering is relative not absolute.

The problem of subjective tolerances in the colour rendering of light sources has been investigated in much greater detail at the National Physical Laboratory (Crawford, 1963). Various objects of both 'natural' and 'artificial' colouring were studied under a light source of variable spectral power distribution. The optical system of the equipment used for the experiment permitted one of a small number of bands of the spectral radiation being withdrawn from the light illuminating the object. This withdrawal could be undertaken slowly or quickly as the experiment demanded. The speed of withdrawal was, in fact, determined in such a way that it could be assumed that the eye would remain completely adapted to the prevailing colour of the light during the whole of the appraisal session. The observer had to note the point at which he could detect a change in the appearance of the objects which he was viewing.

The number of spectral bands chosen by Crawford was the minimum which could adequately describe the performance of an illuminant, and the wavelength limits of the bands were those which were found to produce equal tolerances from the point of view of the amount of excess or deficit of power. These were as follows:

Colour	*Limits of Spectral Band* (nm)
Violet-blue	400–455
Blue-green	455–510
Green	510–540
Yellow	540–590
Yellow-orange	590–620
Red	620–760

Observers made judgments of the tolerances permissible for the difference in energy in the various bands singly or in various combinations. Crawford found that four general rules covered these tolerances:

1. The tolerance for a band is the same whether it refers to an excess or a deficit relative to the reference standard, if expressed as a proportion or percentage of the standard.
2. Bands separated by not less than the width of another band behave as if independent of one another. It would seem that the observer is simultaneously conscious of changes in colour appearance due to all the separated bands at once.
3. Any pair of contiguous bands are subject to half the tolerance of either band alone if they are both deficient or both in excess.
4. If the bands of a contiguous pair vary in opposite directions, however, they behave independently as in (1) and are both subject to the normal tolerance.

In order that the visual effect of the six bands should be correctly related, Crawford proposed that the measured energy should be multiplied by the CIE standard spectral luminosity function, and when comparing two sources, the band luminances so obtained should be adjusted so that the total luminance is the same for each source.

The tolerances permissible to satisfy 95% of the population for most colour rendering purposes were found to be $\pm 10\%$ for single bands and $\pm 5\%$ for the adjacent bands varying in the same direction. Some objects such as the human complexion and certain foodstuffs were found to be slightly more critical in respect of the tolerance permissible in particular bands. Crawford therefore proposed that for 'special cases' the tolerances could be made somewhat tighter, particularly in the red band, where a tolerance of $\pm 4\%$ might be the acceptable limit.

These tolerances, of course, represent change from some reference illuminant. In the study at the Building Research Station the reference illuminant was average daylight. Crawford, however, was mainly concerned with assessing the colour rendering properties of fluorescent lamps made to give approximately white light of correlated colour temperature ranging from 2800 K to 7000 K. He therefore related the colour rendering property of each near-white source to that of the black-body radiator at the same correlated colour temperature. In order to satisfy the demand for a single number which would enable sources to be compared, one with another, for colour rendering quality, Crawford proposed that the amounts of energy in excess or deficit of the permissible tolerance in each band should be summed. This sum could then be used by itself as a figure of merit, increasing with increasing departure from satisfaction, or alternatively it could be subtracted from the value given by the worst possible source (a monochromatic source such as a sodium lamp) to give a figure which would increase with increasing acceptability of colour rendering. In order to simplify the quantification still further, Crawford proposed a classification of lamps in six classes from A to F in descending order of colour rendering quality relative to black-body radiation. His classification is tabulated below:

Range of Figures of Merit	*Colour Rendering Class*
914	A
913–857	B
856–800	C
799–685	D
684–457	E
456– 0	F

Such a figure of merit or classification can strictly be used only to compare two lamps giving approximately the same colour of light. It can, for example, compare the colour rendering given by a 'warm white' fluorescent lamp and an incandescent filament lamp of the same correlated colour temperature since the latter approaches very closely the colour rendering characteristics of the black-body radiator of the same correlated colour temperature. On the other hand, Crawford's classification cannot indicate whether a 'warm white' fluorescent lamp is preferable for a given purpose to a 'north light' lamp of the same figure of merit, not can it indicate whether natural daylight is better than an incandescent filament lamp for a particular colour rendering purpose, nor indeed whether a filament lamp at a correlated colour temperature of 2400 K is preferable to one running at 2900 K. Consequently the lighting consultant, advising on the design of lighting for a clothing or textile shop, and knowing from experience that if the artificial lighting is unsatisfactory shoppers will insist on taking merchandise outside to assess its colour qualities under natural daylight, finds himself in no better position with Crawford's classification than he was before. This difficulty is also one which is encountered in the colour shift method of colour rendering specification proposed by the Illuminating Engineering Society of America (Nickerson 1962; Wyszecki 1959) referred to above. This colour shift method is based on the apparent change in chromaticity of a number of selected colour samples when illuminated by the lamp in question as compared with the reference lamp. In theory, any source could be used as standard, but there is no particular validation for the use of any one source, and the system does not appear to be linked to any subjective investigation as to the amount of shift acceptable, nor to the preference of one direction of shift over another (for example, a shift of saturation as compared with a shift of hue). The method does not take into account the change of adaptation with change of illuminant, and the assumption is made that the data on known minimum perceptible colour difference steps (previously determined by MacAdam and Wright under colour matching conditions) could be applied to determine the colour shift tolerances.

Eight Munsell colour samples are used for this method. The calculations of their predicted change in colour appearance are complex and need the aid of a computer. When these changes have been determined, the problem still remains as to how to express them as figures of merit. Moreover, the final data is only distantly related to the main feature of concern to the lamp manufacturer, which is the spectral power distribution of the lamp. Although the colour shift method has found a certain favour, for example, with the Colour Rendering Committee of the CIE (1965), and also in the Netherlands, the feeling in Great Britain is that its complexities are a high price to pay for the minimal advantages, if any, over Crawford's method. In addition, none of these methods really tell the practising lighting designer what he needs to know. For the moment, unfortunately, he still has to rely very largely on experience.

Special Problems—Hospitals

An example of the above dilemma is the problem of fluorescent lighting in hospitals. Certain clinical tasks which have to be undertaken in hospitals require the use of lamps with colour rendering properties which aid and certainly do not hinder the diagnostic skills of the clinician. Neither Crawford's method nor the colour shift method of colour rendering specification enable a prediction to be made of the best form of artificial light source to be used for the lighting of the diagnostic areas of hospitals. However, it is certain that these diagnostic skills are learned primarily under natural daylight and so it would seem reasonable to assume that the nearest match to natural daylight will prove most acceptable for hospital lighting. This, however, need not necessarily be the case. A knowledge of colour rendering theory leads to the conclusion that diagnosis can, in fact, be aided by the use of lamps which deliberately distort colour rendering as compared with daylight in order to accentuate subtle colour differences. Jaundice, for example, is recognised by a change in the colour of the skin which is more readily detectable under light with a high blue content (light from the clear blue sky, for example) as compared with average daylight. The fluorescent lamp of a very high correlated colour temperature might very well prove more acceptable than a lamp deliberately designed to match average daylight.

Until recently there has been considerable resistance among medical and nursing staff to the use of fluorescent lighting in hospitals on the grounds that the commercially available lamps all cause too much colour distortion for the condition of the patient to be judged easily and reliably. With the advent of 'compact planning' of hospitals, in which some of the service and ancillary rooms are placed in the interior of the building without access to daylight (Chapter 11), it becomes of particular importance to keep the electrical loading as low as possible. This argues in favour of the fluorescent lamp with its triple or even quadruple advantage in luminous efficacy over the incandescent filament lamp.

The consequent need to change over to fluorescent lighting for hospitals, if such lighting could be made acceptable to medical staff, resulted in an investigation conducted by the Joint Committee on Lighting and Vision of the Medical Research Council and the Building Research Board into the characteristics of the spectral power distribution of lighting to give the most acceptable colour rendering of light sources for clinical purposes. The investigation also had as a secondary purpose to see whether a type of fluorescent lamp could be found which had colour rendering characteristics as acceptable for clinical purposes as the tungsten filament lamp. This investigation is described in detail in a Medical Research Council Memorandum (1965).

The experiments consisted of making subjective judgments, by skilled staff, on various critical clinical tasks, using different types of illuminant. Among the ninety observers who took part in the experiment were bacteriologists, physicians, nursing staff, medical laboratory technicians, pathologists, dermatologists, radiologists, anaesthetists, consultants, surgeons, ophthalmologists, dentists, medical artists and physicists. Observers were each asked to make the kind of judgments which were characteristic of the course of their normal work, among the objects studied being pathological samples, blood samples with oxyhaemoglobin, reduced haemoglobin, etc., bacteriological cultures, artificial teeth, and test reactions such as Lange's colloidal gold reaction which involves delicate changes of colour from pale red to pink and bluish pink. In addition to these objects, patients with different conditions were examined, including patients with dermatological lesions or skin conditions, cases with cyanosis and early cases of jaundice. In addition to all these visual tasks, printed colour charts were also observed in each test to serve as a cross-check between observers and between different tests.

The light sources to be tested were arranged in a special fitting which enabled any one of four different light sources to illuminate the visual task. The illumination was adjusted so that under every source the task was illuminated to 25 lm/ft^2 (270 lux). A particularly important feature of the experiment was that the lamps were screened to prevent a direct view by the observer, so that he was unaware of the nature of the light source, and in particular was unaware when fluorescent lighting as opposed to filament lighting was being used.

The sources used for the investigation are given in Table 7.2. The table also shows the correlated colour temperature and the colour rendering classification as determined by Crawford's NPL method. (The classification B/C for lamps numbers 8 and 10 is an indication that some lamps of a batch of these types fell just within class B and some just within class C.)

Table 7.2

SOURCES TESTED FOR CLINICAL USE (FROM MEDICAL RESEARCH COUNCIL 1965)

Reference No.	*Make*	*Type*	*Correlated colour temperature* (K)	*NPL colour-rendering classification*
1	—	Tungsten	2400	A
2	—	Tungsten	2650	A
3	—	Tungsten	2800	A
4	'Colour 32' (Philips)	Fluorescent	2900	C
5	'Hospital' (Atlas)	Fluorescent	3600	D
6	'De Luxe Natural' (Atlas)	Fluorescent	3700	D
7	'Colour 34', old type (Philips)	Fluorescent	3800	B
8	'*Kolorite*' (Mazda)	Fluorescent	4000	B/C
9	—	Mixture of blue fluorescent and tungsten	4100	A
10	'Colour 34', new type (Philips)	Fluorescent	4100	B/C
11	'Daylight' (Osram)	Fluorescent	4100	F
12	'Super 5 (Natural 3)' (Atlas)	Fluorescent	4200	D
13	'Cool White' (Philips)	Flourescent	4300	F
14	'Colour Matching' (Osram)	Fluorescent	5000	B
15	'Colour 55' (Philips)	Fluorescent	6500	B

Lamps 1, 2 and 3 in the table are tungsten filament lamps which were chosen to cover a range of correlated colour temperature, while lamp number 4 is a fluorescent lamp which is a good imitation of the chromaticity of tungsten filament light at 2900 K. Lamps 5 to 13 are those in the mid-range of correlated colour temperature, and emit light corresponding in colour to that of warm phases of daylight. They have a range of colour rendering classification from A (for the special mixture number 9) to F (for the 'daylight' or 'cool white' colours). The nearest lamp to the full black-body radiator (with the exception of the tungsten filament lamps) is the mixture of one 80 W blue fluorescent lamp with four tungsten 60 W line filament lamps running at a correlated colour temperature of 2460 K. The other lamps in this group have progressively greater deficiency of light in the red and the blue-green regions of the spectrum, coupled with excess in the far blue and green-yellow regions, as the colour rendering classification worsens.

The tests were carried out with four different lamps in each series; in the first two series it was planned to take cross-sections of the gamut of lamps available in two different ways. First, a group of lamps, all of about the same correlated colour temperature (3800 to 4100 K), covering the range of colour classification from A (the special mixture number 9, having approximately full-radiator type spectral power distribution), through B (Philips Colour 34 old series), B/C (Mazda *Kolorite*), to F ('daylight'). The second series was a group of lamps having good colour rendering in terms of the NPL classification (A to B) but a range of correlated colour temperature from 2650 K (tungsten filament lamp) through 3800 K (Philips colour 34 old series) and 5000°K ('colour matching' type) to 6500 K (Philips colour 55).

After the first two series, three further series of tests were carried out, one to check the performance of correlated colour temperature in the range 2400 K to 3800 K (with lamps with fairly good colour rendering classification), one to check the effect of colour rendering classification with a group of lamps (classified B to D) with correlated colour temperatures between 3700 K and 4000 K, and the third series to repeat the first series but with the special mixture lamp replaced by a commercial lamp with poor colour rendering classification, and with the Mazda *Kolorite* lamp replaced by a lamp with enhanced emission of red light and a correlated colour temperature of 3600 K, which had been put forward by a manufacturer as a possible solution to hospital requirements.

In each test each of the lamps was compared against one other in a direct pair comparison test. The observer's task was merely to look at the object or the patient and to say whether the lamp gave better or worse colour rendering than the one immediately preceding. Consequently no ranking was required of the observers themselves, and the task was a relatively easy and quick test to carry out. The lamps were presented to the observer in a series such as A B C D B A C A D C B D A. Thus each lamp appeared immediately before as well as immediately after every other lamp so that the comparison was carried out twice in every test but in reverse order. Each test for each observer was therefore considered as two 'observer occasions'.

The statistical analysis of the results was carried out by means of a matrix in which each preference was represented by a score of one and each occasion on which another lamp was perferred by a score of zero. The scores were summed, and tests were performed for consistency, any inconsistent results of the type A better than B, B better than C, C better than A, being neglected in the scoring. (This is the procedure adopted as standard by statistical processers, but it should be noted that in this type of experiment 'inconsistency' does not necessarily imply an erroneous or unsatisfactory observation, but could equally be an indication that a barely perceptible difference between the colour rendering properties of the sources actually exists.)

The actual scores and their statistical significance are given in the Medical Research Council Memorandum. The results may be summarised as follows:

1. The order of preference was the same for all the examination tasks, with the exception of that for early jaundice, which will be considered separately.
2. The closer the spectral power distribution approached that of a full radiator, the higher was the preference. The special mixture source was easily the first preference, with the Philips Colour 34 (old series) lamp a

clear lead over others. This Colour 34 lamp was a practical commercial source with the nearest approach to the full radiator spectral power distribution, and since it appeared at the time to be a readily obtainable light source whereas the special mixture was not, it was used in place of the mixture source in all subsequent tests.

3. A source of correlated colour temperature in the region of 3800 K was preferred to sources of either higher or lower correlated colour temperature, the preference apparently decreasing as the correlated colour temperature increased, and also as it decreased from 3800 K to that of the tungsten filament lamp.
4. The preference decreased steadily as the correlated colour temperature was lowered over the above range, and even further to a tungsten filament lamp with a correlated colour temperature 2400 K.
5. Over a smaller range of colour rendering classification, preferences were in the order of this classification.
6. Sources of a correlated colour temperature in the region of 4100 to 4300 K, even if of poor colour rendering quality, were preferred to a source of correlated colour temperature 3600 K, even if the latter had better colour rendering quality.
7. Cases of early jaundice were recognised more readily if the light source had an output higher in the blue region of the spectrum (or was relatively deficient in red). The complete series of results on the examination of early jaundice, which is a special problem, showed a certain lack of precision in the specification of the optimum type of lamp for this purpose. The best lamp for all other purposes (Colour 34) appeared to be reasonably satisfactory for the detection of early jaundice as well. The Medical Research Council therefore recorded a decision that there is no justification for modifying the recommendations for general hospital lighting in clinical areas on the score of jaundice detection. This does not rule out the possibility that a special examination lamp for this purpose might be provided in suitable circumstances.

This series of experiments has led to the specification by the Ministry of Health* of a lamp for general hospital use in clinical areas having a correlated colour temperature in the region of 4000 K and a colour rendering classification of B or better on the Crawford NPL scale. It was recommended that the target for maximum deviation from a full radiator should not exceed a 'sum of excesses' of more than 30 on the NPL scale. The issue by the Ministry of Health of this specification was intended to achieve a uniform standard of colour rendering and a uniform colour appearance of light throughout all hospitals. At the present time lamp manufacturers are undertaking design of suitable lamps to meet the specification.

At the time of formulating the specification, it could be met, at the cost of low efficacy, by a commercial lamp (the Philips Colour 34, old series). Im-

* Now Department of Health and Social Security.

provements have already been made. Philips' Colour 37 closely approaches the target for colour rendering properties, and has been recommended by the Ministry of Health for use in clinical areas in hospitals. Other manufacturers are endeavouring to meet this specification without too great a loss in efficacy.

Colour Rendering and Illumination Level

Studies of preferred colour rendering quality, such as the study of lighting in hospitals described above, give rise to a difficulty which is not yet resolved. The preferred lighting is of lower luminous efficacy than that of lighting with poorer, less-preferred colour rendering qualities, and it must therefore cost more to provide the preferred amenity *unless* it can be shown that less light of the preferred quality leads to equal visual performance or equal visual satisfaction.

Uncompleted work by Hopkinson, Loe and Rowlands (1969) shows that the advantages, as regards visual performance, of the higher illumination obtained for the same power per unit area by light from fluorescent tubes of a higher efficacy but lower colour rendering quality are marginal and are barely detectable by a sophisticated technique. This is to be expected from the known nature of the performance versus illumination relationship (Chapter 3) at moderate to high levels of illumination. However, it has not been deliberately studied and reported upon before. On a basis of visual performance and colour preference alone, therefore, this work indicates that in situations where preferences are strong for light of good colour rendering, it is better to provide this light at the lower level of illumination which results from the same power per unit illuminated area, than to reject the better quality light on grounds of cost.

The situation is not simple, however. Certain types of light distribution in rooms are very sensitive to the absolute level of lighting and in some circumstances the lower level of lighting which results from the use of lamps of lower efficacy may give rise to a less satisfactory overall effect. This is particularly the case where the artificial lighting is used as a supplement to daylight where the sky is visible through the window (Chapter 11). A great deal of further work is required before any firm pronouncement can be made on the complexities of the problem.

References

Bodmann, H. W., G. Söllner, and E. Voit (1964): 'Bewertung von Beleuchtungsniveaus bei Verschiedenen Lichtarten.' Proc. CIE 15th Session (Vienna, 1963). Vol. C, 502. Commission Internationale de l'Éclairage, Paris.

Bodmann, H. W. (1967): 'Quality of Interior Lighting Based on Luminance.' *Trans. Illum. Engng. Soc.* (*London*) **32**, 22–40.

British Standard 950:1967. 'Artificial Daylight for the Assessment of Colour: Part 1. Illuminant for Colour Matching and Colour Appraisal.'

British Standard 1853:1967: 'Tubular Fluorescent Lamps for General Lighting Service.'

British Standard 2600:1961: 'Colours for Building and Decorative Paints.'

CIE (1965): 'Methods of Measuring and Specifying Colour Rendering Properties of Light Sources.' Commission Internationale de l'Éclairage, Publication No. 13. Paris.
Collins, J. F. (1965): 'The Colour Temperature of Daylight.' *Brit. J. Appl. Phys.*, **16**, 527–532.
Condit, H. R. and F. Grum (1964): 'Spectral Energy Distribution.' *J. Opt. Soc. Amer.*, **54**, 937–944.
Crawford, B. H. (1963): 'The Colour Rendering Properties of Illuminants: The Application of Psychophysical Measurement to their Evaluation.' *Brit. J. Appl. Phys.*, **14**, 319–328.
Gloag, H. L. (1961): 'Colouring in Factories.' DSIR Factory Building Study No. 8. HMSO, London.
Henderson, S. T. and D. Hodgkiss (1963, 1964): 'The Spectral Energy Distribution of Daylight.' *Brit. J. Appl. Phys.*, **14**, 125–131; **15**, 947–952.
Hopkinson, R. G., Loe, D. M. and Rowlands, E. (1969): Performance, Preference, and Colour of Fluorescent Lighting. University College Environmental Research Group Note No. 1969/3, London.
Kelly, K. L. (1963): 'Lines of Correlated Colour Temperature Based on MacAdam's (U, V) Uniform Chromaticity Transformation of the CIE Diagram.' *J. Opt. Soc. Amer.*, **53**, 999–1002.
Kruithof, A. A. (1941): 'Tubular Luminescence Lamps for General Illumination.' *Philips Tech. Rev.*, **6**, 65–73.
Medical Research Council (1965): Memorandum No. 43. 'Spectral Requirements of Light Sources for Clinical Purposes.' HMSO, London.
Nickerson, D. (1962): 'Interim Method of Measuring and Specifying Colour Rendering of Light Sources.' *Illum. Engng.* (*New York*), **67**, 471–495.
Wright, W. D. (1964): 'The Measurement of Colour.' Adam Hilger, London.
Wyszecki, G. (1959): Proc. CIE 14th Session (Bruxelles, 1959), 146. Commission Internationale de l'Éclairage, Paris.

8

Apparent Brightness and Adaptation: Attention and Distraction

The Apparent Brightness Concept and its Evaluation

The need for the quantitative evaluation of human sensations is felt in all branches of environmental design. For many years there has been available a subjective scale of loudness which expresses in numerical terms the subjective estimate of loudness magnitude in relation to the physical measure of sound pressure. This subjective scale of loudness has been internationally agreed and acoustic design would be difficult or impossible without it.

The need for a subjective scale of brightness has been felt in lighting for the past thirty years, particularly when the design has to cater for a wide range of conditions such as visibility on the street at night and by day, or in the lighting of an interior by daylight and by artificial lighting. The fact that so far such a scale of 'apparent brightness' has not been internationally adopted is due entirely to the fact that workers in the design of the visual environment tend to be highly specialised and so those who design street lighting, for example, interpret physical photometry on one scale of subjective attributes while those concerned with the lighting of building interiors interpret on a different scale. It is only when one designer has to deal with a whole range of physical conditions that the need for a scale of apparent brightness becomes as acute as the need for the scale of loudness.

A typical example of the need to consider lighting design in terms of apparent brightness is the problem of the lighting of road direction signs. It is essential that these signs should be equally conspicuous by night and by day. A motorway sign consisting of white letters of perhaps 80% reflectance, on a blue background of perhaps 30% reflectance, receiving full sunlight, may well have an average luminance of the order of 3500 cd/m^2 (1000 ft–L) (Fig. 8.1). But to achieve the same degree of conspicuity at night does not require such a high luminance and, in fact, a luminance of 3500 cd/m^2 for an object the size of a motorway sign would be so blindingly brilliant as to cause acute glare disability and so obscure everything in the field of view, especially on the carriageway itself. The same conspicuity will be achieved with a much lower luminance. Because of the lower level of adaptation of the eye on the

Fig. 8.1 *Motorway sign by day*

road at night, the same conspicuity will be achieved if the sign has a luminance of only a small fraction of that prevailing by day (Fig. 8.2). For example, Fig. 8.3 shows that the *apparent brightness* of the signs under the two conditions would be about the same even though the *physical brightness* may differ by 300 : 1.

A valid scale of apparent brightness related to adapting conditions would permit the necessary luminance of the sign in any other prevailing conditions of natural or artificial lighting to be determined. On the other hand, the implementation of the design to give the necessary apparent brightness must, of course, be in terms of physical energy, that is, in terms of luminance. Apparent brightness is therefore a concept which serves as a transfer process. Final design must always be in terms of physical quantities, just as in acoustic design final criteria are in terms of sound pressure level.

The concept of apparent brightness is easier to explain to the layman than

Fig. 8.2 *Motorway sign by night*

to the trained physicist or engineer. Apparent brightness is simply a numerical measure of 'what we see'. When the eye, adapted to one situation (the motorway at night), sees a sign 'as bright as day', that is, of the same apparent brightness as the eye sees it when light-adapted to daylight, even though the photometer may show that the physical brightness of the sign differs by 300 :1, the physicist uses words such as 'the eye deceives'. In his physical terms the eye sees as equal things which are unequal. On the other hand, the layman considers that it is the photometer which deceives, because it measures two things demonstrably and manifestly equal as being vastly different. In the same way, the layman can be given control of a light source on a dimmer, and if he is asked to turn up the dimmer so that the light is now 'twice as bright' as it was before, he will do this without questioning and will give a result which is obviously meaningful to him because if he does this several times, his settings will be consistent about a steady mean. The physicist or

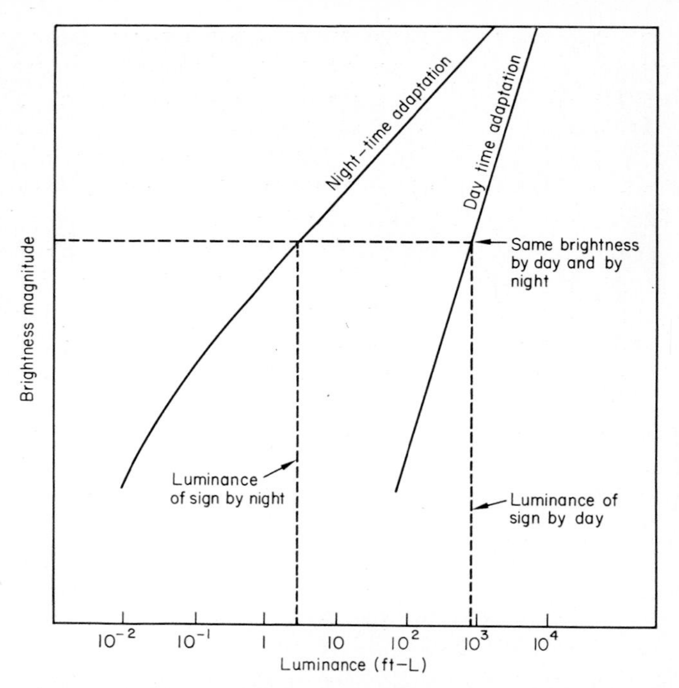

Fig. 8.3 *Apparent brightness magnitude related to luminance and adaptation level*

engineer, on the other hand, is inhibited from making any such setting and will probably ask questions such as 'what do you mean by twice as bright?' or, much more likely, will refuse to make any attempt to do what he considers to be meaningless if not impossible.

The Apparent Brightness Concept

The concept of apparent brightness, as an entity which can be quantified for design, rests on the assumption that everything we see is evaluated, as far as its brightness is concerned, in terms of a reference level which is associated with the state of adaptation of the eye at the time. Consequently if the eye is dark-adapted, a given luminance very much higher than this average level will have a high apparent brightness; on the other hand, if the eye is adapted to a luminance much higher than that of the object, the object will then look dark. The apparent brightness is therefore a function not only of physical luminance of the object itself, but of the physical luminance of the surroundings which govern the adaptation of the eye at the time.

The concept of apparent brightness, so far as the design of lighting and the visual field is concerned, originated in the 1930s and was made specific by several workers of whom Beuttell may have been the first, because his ideas led to a series of experiments by Craik (1938, 1940), who derived what appear to have been the first data relating physical luminance, subjective sensation, and adaptation. Craik's experimental work was based upon a technique of binocular matching to which there are some objections, and it may have been for this reason that his work had little subsequent influence. In parallel, and independent of the Beuttell–Craik work, Hopkinson (1936) was developing the concept of apparent brightness from work originally undertaken to develop a method of photographic representation of lighting installations to give a true subjective sensation of the original. Hopkinson found that the necessary requirement for true photographic representation was that the apparent brightness of the reproduction, as viewed in the adapting conditions of the viewing room, should be the same as the apparent brightness of the original. He established the necessary relations between the luminances of the original and the luminances of the reproduction under the new adapting conditions to predict what should be the required densities of the photographic reproduction for true representation. He derived a series of scales of apparent brightness relating luminance to the subjective sensation by a method entirely different from that of Craik (Hopkinson 1939, 1941 with Waldram and Stevens).

Again, Wright had proposed a scale of subjective brightness which at first did not take adaptation into account, but which subsequently, as a result of an investigation by his collaborator Pitt (1939) gave rise to a series of apparent brightness data using a similar binocular matching technique to that used by Craik. Like the data of Craik, that of Pitt also resulted in no follow-up, and the technology of lighting design based on apparent brightness at the moment derives from Hopkinson's data.

Hopkinson's scales of apparent brightness were applied originally, during the period 1937–1939, to the design of street lighting installations, and subsequently to problems of visiblity of radar displays and other wartime visibility problems. Subsequently the first application of apparent brightness design techniques to interior lighting was made by Hopkinson and his colleagues at the Building Research Station for the design of a reconstruction of rooms in the National Gallery, where the requirement was to realise certain concepts of the architect for brightness emphasis on the pictures (Hopkinson 1948, 1951). Later Waldram (1954) incorporated Hopkinson's apparent brightness data into a comprehensive method of lighting design for appearance particularly with reference to lighting schemes which require special emphasis on specific objects, problems which are found in churches and cathedrals, art galleries, exhibition halls and other places where light is used as a means for creating special effects. Waldram has worked out a detailed technique for the lighting engineer to employ.

There are two basic requisites for the operation of the apparent brightness concept in terms of quantitative lighting design. First, there is a need for valid numerical scales of apparent brightness. Because of the profound influence which the adaptation conditions have on the subjective evaluation of brightness, scales have to be specified in terms not only of the physical luminance of any particular object or area in the visual field, but also in terms of the adapting conditions. These adapting conditions must also be expressed in physical luminance, and so the second requirement is that there must be a valid method of determining the adaptation brightness in terms of physical luminance, under the prevailing conditions.

The scales of apparent brightness put forward by Hopkinson (see Fig. 8.4) have been used by many workers for the design of lighting under a wide range of conditions. The scales were obtained by a method called by Hopkinson the 'contrast ladder method', that is, judgments of steps of equal suprathreshold contrast under controlled conditions. Observers saw two patches in an otherwise uniform field, the brightness of one patch being noticeably greater than that of the other (actually by about five 'just noticeable difference' steps). The observer was then asked to adjust the brightness of the darker patch until it was now brighter than the other patch by the same amount as before. This was repeated so that a series of steps of equal brightness difference, as judged

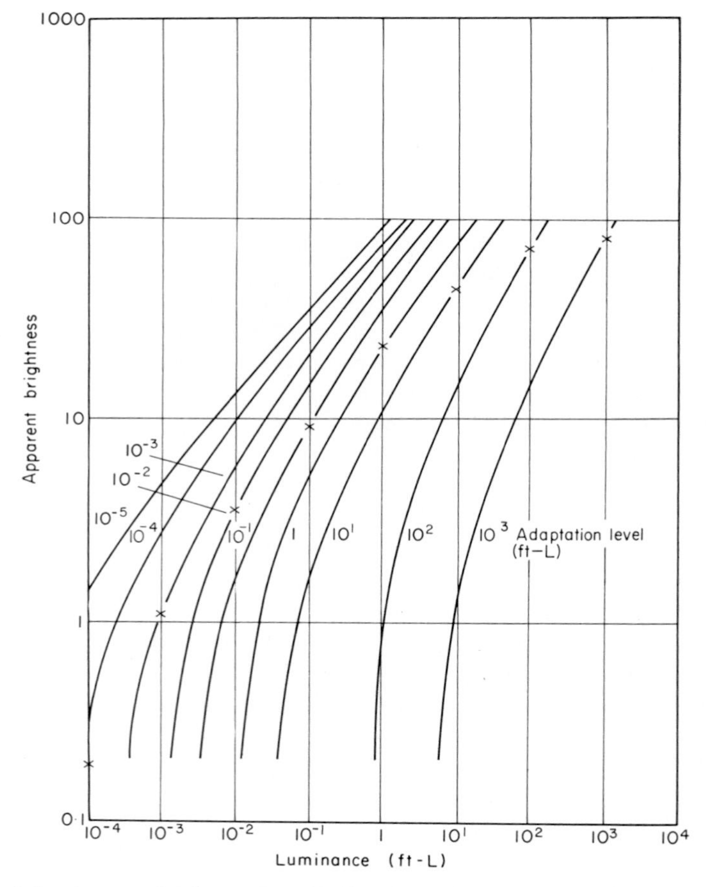

Fig. 8.4 *Apparent brightness data (Hopkinson* et al. *1941) plotted on log/log scale*

subjectively, were determined and these could be related to the corresponding physical luminance. An arbitrary scale of equal steps of apparent brightness could then be set up to relate with the corresponding physical luminance to give a relationship which corresponded to the adaptation level of the experiment. Similar contrast ladder determinations were made at other adaptation levels, and the series of curves shown in Fig. 8.4 represent the best fit to the data.

This scale was found to work tolerably well in practice, particularly at low levels such as those experienced in street lighting, in radar control rooms, and in some examples of the interior artificial lighting of buildings. Subsequently, Hopkinson found that there were some discrepancies when the data were applied to daylighting conditions. For this reason some attempts were made by Hopkinson (1951, 1957) to establish a new set of data by different methods of subjective evaluation. In particular, a technique of direct estimation of brightness magnitude was evolved from a method which had been developed by Stevens (1955 *et seq.*) for the estimation of the loudness of sound and upon which the internationally agreed sone scale of loudness assessment is in the first instance based. This brightness magnitude estimation data proved to be more reliable at high levels, but unfortunately by the time that it became known, other workers had already advanced some way in the development of lighting design technologies based on the earlier data. An attempt by Hopkinson in 1959 to obtain international agreement for apparent brightness scales based on the method of direct magnitude estimations was not immediately followed up (Hopkinson 1960).

The two sets of data, although obtained by different methods, can, in fact, be shown to be in tolerable agreement. This is rather remarkable in view of the fact that there are always difficulties encountered by observers in making subjective judgments of brightness by any method. The two sets of data are incorporated in Fig. 8.5, in which two subjective scales have been related by

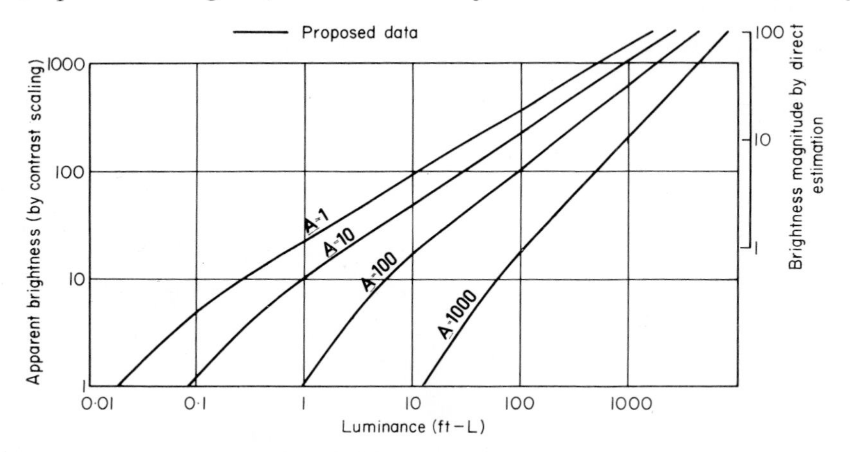

Fig. 8.5 *Proposed apparent brightness relationship to combine contrast scaling data and brightness magnitude data*

A = Adaptation level in ft-L

a constant ratio such that the best fit of the curves is obtained. This exercise permits a set of postulated data, not themselves based directly on experimental results, but derived from the two independent sets of experimental results of the contrast ladder method and the method of direct brightness magnitude estimation.

It may be questioned why such a synthesis of the two sets of data is necessary rather than a decision to employ the one or the other as was suggested by Hopkinson in 1959. It may be worth studying the situation. The contrast scaling data have the following advantages:

1. The scales have now been fairly widely used, and workers, particularly Waldram and his associates, who have used them are familiar with the subjective attributes of the numerical values which they give. It would therefore set back the discipline to some extent to put forward a new set of scales with unfamiliar values.
2. Information obtained by the use of these scales appears to accord with practical experience over the range for which they have been employed, particularly for moderate and low-level luminance situations.
3. The unit of the arbitrary scale is of the order of one just noticeable difference under practical conditions. This is not an important consideration because the just noticeable difference is profoundly affected by a great many factors such as the size of the area under observation, the position in the field of view, the nature of the immediate surroundings and so on. However, it is an advantage which has sometimes been quoted and may therefore be worth recording.

The direct magnitude estimation data have the following advantages:

1. The scale of brightness magnitude accords well with subjective sensation, in that a magnitude of 4 appears 'twice as bright' as a magnitude of 2: a magnitude of 20 appears 'twice as bright' as a magnitude of 10; and so on.
2. The numerals used in the brightness estimations are, on average, the numbers actually chosen by people to express their subjective impressions. Thus a luminance of 100 ft–L (350 cd/m^2) seen in an adaptation of 10 ft–L (35 cd/m^2) will, on average, be given a number of the order of 10 when different observers are asked to assess its magnitude. (The contrast scaling data calls for the use of numerals which are very much greater than those which people appear ready to use when asked for direct magnitude estimations.)

This latter property of the brightness magnitude estimation data may be important. An experiment was once conducted (Hopkinson 1960) to induce observers to use high numerals of the order required by the contrast-scaling data, and it was found that people were to a great extent confused by this in their attempts to make magnitude estimations. Consequently, in any attempt to synthesise the two methods of subjective brightness assessment, it must be appreciated that the scale of apparent brightness deduced from the contrast-scaling experiment is entirely arbitrary, whereas the scale of direct magnitude estimation is not arbitrary since these numbers were given by the observers (on average) to the brightness magnitude as they saw it during the experiment. Such a scale has therefore some absolute meaning over and above its arbitrary character.

The postulated data of Fig. 8.5 were adjusted to give a good fit to the original contrast-scaling data at low levels of luminance, and to the brightness magnitude estimations at high levels of luminance. Two scales are actually given on the figure with the intention that those who are familiar with the original data can continue to use values which correspond closely to these original values, while on the other hand it may be found that the newer scale, based on the brightness magnitude estimation, may prove more useful when discussing lighting designs with others, because these values can be expected to accord more closely with the layman's own estimate of the magnitude of brightness as he sees it.

The Adaptation Luminance

While the determination of the apparent brightness of a given object in terms of its physical luminance, and of the adaptation conditions, is a straightforward procedure using data such as that shown in Fig. 8.5, the success of designs based on the data depends upon the ability to specify the adaptation level and therefore the appropriate curve to be selected. In the experimental studies upon which the data are based, Hopkinson took as the adaptation level the uniform luminance of the surround field. In practice, therefore, the *adaptation level* should be defined as *the luminance which results in the object having the same apparent brightness as it would have in a uniform field of that luminance*. The visual field, however, rarely if ever has a uniform luminance, so that the apparent brightness of any area is affected not only by the average luminance of the whole field, but also possibly by simultaneous contrast with the luminance of contiguous areas. Hopkinson (1965) has recommended that the adaptation luminance should be taken as the numerical equivalent in foot-lamberts of the vertical illumination expressed in lumens per square foot at the observer's eye, or in metric terms, as the numerical equivalent in apostilbs of the vertical illumination expressed in lux. This is an expedient which is known to work satisfactorily when the range of luminance, including that of the light sources, is restricted to within a moderate range, about 10:1, of luminance above or below the adaptation level.

One of the most important steps necessary before the apparent brightness concept can be used extensively in lighting design is to specify precisely the way in which the adaptation luminance in any given situation should be defined. It may well prove to be that Hopkinson's suggestion (the luminance equivalent of the illumination on the eye) is adequate for all practical purposes.

On the other hand, Waldram believes that apparent brightness is determined essentially by the local field, within 6° of the object of regard. Further experimental work to determine adaptation specification is therefore necessary.

Limitations in the Apparent Brightness Concept

While the use of the apparent brightness concept has achieved some success, particularly in the hands of Waldram, it has been criticised on various grounds, one of which is that it is unnecessarily complex, and that a simplified concept will serve the same purpose. As long ago as 1941 Wright put forward a 'brill' scale for expressing subjective brightness on a single scale in which adaptation is not included, and this in modified form has been revived by Jay (1967) and others. Indeed, anyone who wishes to search for reasons for denying the validity of the apparent brightness concept need not go very far. Two difficulties which assert themselves are those associated with the phenomena of simultaneous contrast and brightness constancy. In practice, the apparent brightness of an area is affected not only by the general adaptation level, but also quite markedly by the luminance of neighbouring areas. If the neighbouring areas differ considerably from the luminance of the object under study, the effect on the apparent brightness can be very noticeable. This effect is called *simultaneous contrast*, and it is not possible at the present stage of knowledge to offer analytical methods to handle it in apparent brightness design. In most practical situations it is felt not to be of great importance, and the experienced practitioner soon learns to make arbitrary allowances to take it into account. Nevertheless it is very simple to devise a demonstration employing simultaneous contrast in order to lay claim to demolish the apparent brightness concept. Indeed, Marsden (1969) suggests that induction between all luminances in the visual field can explain the apparent brightness effect.

The effect of brightness constancy can cause more serious difficulty. *Brightness constancy* is the phenomenon by which the eye appears to be able to differentiate between a light surface poorly illuminated and a dark surface which receives high illumination, even though the two surfaces may have the same physical luminance. The eye can distinguish the separate components of reflectance and illumination only when the source of light and its position relative to the surface being viewed can be seen. For example, a white ceiling lit by daylight from a side window appears white all over its entire surface, even though its luminance adjacent to the window may be five times that in the remote part of the room. In fact, the ceiling may look almost uniformly bright and white, until the observer is alerted to the situation and studies it with greater thought. Consequently in making a prescription of apparent brightness, the architect might well be influenced profoundly in his numeration by such brightness constancy effects, and if these were not allowed for in the design by a percipient designer, could lead to some degree of failure. Here again practical experience is the only present substitute for an analytical procedure, and Waldram claims that constancy has little effect. Yet it was one of the overriding difficulties which Hopkinson experienced and which led (1951) to new experiments to check the original apparent brightness data. Lynes (1968) and Jay (1967, 1968) have further drawn attention to the effects of constancy in practical lighting situations.

Apparent Brightness—An Attribute of Brightness Sensation

It must further be borne in mind, and cannot be stated too often, that the apparent brightness concept is nothing more than a method of stating numerically the magnitude of brightness sensation. It is related to the perception of brightness and that is all that it sets out to do. It does not correlate with any affective state. It cannot necessarily correlate with pleasure, discomfort, distraction, emphasis, or any other similar attribute which stems from other areas of the consciousness than those which govern perception of brightness magnitude alone. It often happens that misunderstanding of the apparent brightness concept is due to failure to realise that this is the case.

Hopkinson and Bradley (1959) designed an experiment in which subjects were asked to adjust conditions in a model room with glaring light sources until the sensation of discomfort was 'twice', 'ten times', 'one-half' and 'one-tenth' of the original level of discomfort. Although the observers were being asked to express a secondary sensation (discomfort) in numerical terms, the results were consistent and showed no excessive variance. What was significant in the present context, however, was that the estimations of discomfort magnitude did not correlate directly with estimations of apparent brightness. Whereas under the given conditions of adaptation apparent brightness was shown to be related to luminance by the relationship

$$M = kL^{0\cdot5}$$

where M is apparent brightness magnitude and L is luminance, the magnitude G assigned to the glare discomfort was found under the same conditions to be expressed by:

$$G = KL^{1\cdot23}$$

In other words, whereas the luminance must be raised four times to appear *twice as bright*, the luminance must be raised only 1·8 times to appear *twice as uncomfortable.*

This is an important finding, because it demonstrates that even though a factor like glare discomfort may be correlated with, among other things, luminance of the glaring source, it does not follow that this correlation will be directly related to the primary sensation of brightness. If this is true of a sensation such as glare discomfort, which is not too far remote from the primary sensation of brightness, it must be even more true for attributes such as emphasis. There is a wide open field of psychophysical research waiting for investigation.

The Application of the Apparent Brightness Concept in Lighting Design

The stages necessary in the use of the apparent brightness concept in lighting design are logical. The object is to realise in terms of subjective effect the requirement prescribed by the designer. For example, if the lighting engineer has an assignment to light a supermarket so that it satisfies the requirement that it should look 'as bright as day', it is not necessary to supply the full illumination of outdoor daylight. The first requirement is to establish the value of apparent brightness which will satisfy the client that the supermarket does, in fact, *appear* as bright as day, and then to supply this *apparent brightness* by whatever level of lighting which proves to be necessary when taking into account the particular adaptation conditions in the supermarket.

The stages of design are as follows:

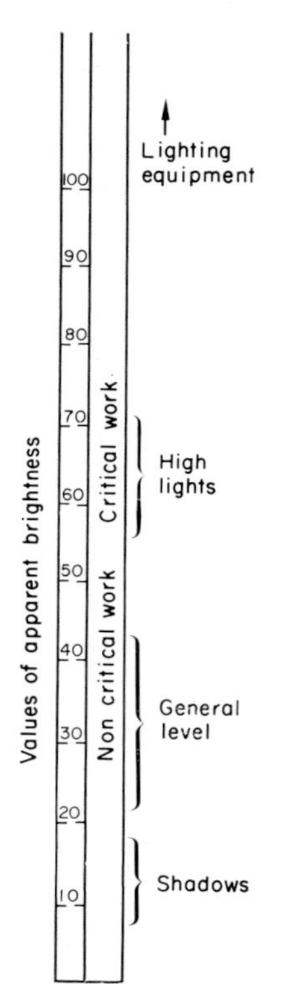

Fig. 8.6 *Scale of apparent brightness and some typical values for artificially lighted interiors (From Waldram 1954)*

1. Prescribe the necessary apparent brightness of the significant areas of the installation in numerical terms, using one or other of the scales given in Fig. 8.5. Waldram has drawn up a scale of visual attributes to correlate with Hopkinson's original scale of apparent brightness. Waldram's correlation is shown in Fig. 8.6.
2. A datum will have to be selected for the calculation. This datum may either be arbitrary, or it may be determined from external considerations such as, for example, the amount of illumination which is necessary to enable workers to see to do a particular visual task accurately. This necessary illumination will then determine the luminance of this particular part of the visual field. In a concert hall, for example, the datum luminance might be determined by the amount of light necessary on the orchestra stands to enable the musicians to see to read their music, as this is probably the only critical visual situation in the hall.
3. Relate this luminance to the apparent brightness requirements, and thence find the necessary adaptation condition for this relation to hold good.
4. Find on the scale of apparent brightness appropriate to the adaptation conditions the necessary values of luminance which will give the required apparent brightness values of all the other selected areas in the field of view.
5. Thence calculate the distribution of light to enable these luminances and hence the required apparent brightnesses, to be achieved.

The stages in the calculation, although straightforward

once the basis of the apparent brightness concept has been mastered, are not easy to grasp in the initial stage. As a very simple example, consider the lighting of an exhibit in an exhibition hall where it is known from other considerations (e.g. the architect's concepts of emphasis, etc.) that the apparent brightness of the exhibit should be 100 and that this should be seen against a background of apparent brightness 10. The exhibit is of such a nature that, in order that people may be able to comprehend all the details of the exhibit, an illumination of not less than 500 lux (or, say 50 lm/ft^2) should be provided, while the average reflectance of the exhibit is 20%. The average luminance of the exhibit in the units used in Fig. 8.7 is therefore $0{\cdot}2 \times 50 = 10$ ft–L. This is therefore the datum point referred to in (2) above. The procedure is then as follows:

First, find the adaptation level at which a luminance of 10 ft–L has an apparent brightness of 100 as prescribed. Reference to Fig. 8.7 shows that this adaptation level is approximately 1 ft–L.

It is now required to find the luminance of the surround to the exhibit which must have an apparent brightness of 10. By following the curve for Adaptation = 1 ft–L on Fig. 8.7, it is seen that an apparent brightness of 10 corresponds to a luminance of 0·25 ft–L. This is the required luminance of the surround.

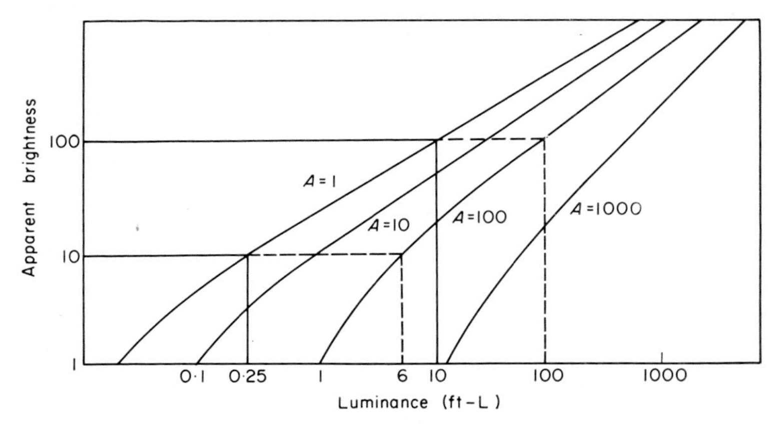

Fig. 8.7 *Apparent brightness (working example)*
A = adaptation level (ft–L)

The final stage is then to provide lighting on the surround which, taken with the reflectance of the surround, will provide a luminance of 0·25 ft–L. Thus if the reflectance of the surround is 10%, this would be achieved by a value of illumination of 2·5 lm/ft^2 (or about 27 lux).

Sometimes the problem is posed in a different way. It might happen, for example, that due to circumstances beyond the control of the designer, such as a change of interior illumination due to variable daylight, the adaptation conditions might be 100 ft–L instead of 1 ft–L. Reference to the appropriate adaptation curve (for 100 ft–L) shows that the desired apparent brightness of 100 is given by a luminance of 100 ft–L. To achieve this luminance, however, the illumination of the exhibit would need to be 100/0·20 = 500 lm/ft^2 since the reflectance of the exhibit is only 20%. In the same way, the surround luminance, to give an apparent brightness of 10, would be seen (from Fig. 8.7), by tracing along the 100 ft–L adaptation curve, to be 6 ft–L. Note that at this higher level of adaptation, this required apparent brightness difference results from a luminance ratio of 100:6 = 16:1 (approx.) whereas at the lower adaptation level, the same apparent brightness difference is given by 10:0·25 = 40:1.

In practice, most designs which find it necessary to employ the apparent brightness concept are more complicated than the simple case given above. A lighting specialist who devotes attention to the method can, however, undoubtedly achieve success in its application, but one of the greatest difficulties facing the designer at the moment is that of obtaining a satisfactory prescription of the desired lighting effect in terms of apparent brightness. The project leader, who may be an architect, may not be familiar with the subjective attributes of the apparent brightness scale, and although he can be helped by information such as that of Fig. 8.6, it is more than likely that the lighting designer may have to write his own apparent brightness prescription. This is, in fact, what Waldram does in his work, and so he himself is to some extent the designer as well as the engineer.

There is one possible way of overcoming this major disability. Sometimes he can ask the project leader or architect to demonstrate some similar scene to the one which has to be designed and to describe his requirements in terms of this other scene. For example, an architect designing a concert hall might ask the lighting designer to arrange that the concert platform as seen by the audience should stand out boldly in brightness over the rest of the interior, while the walls and ceiling of the auditorium should be unobtrusive. These words by themselves mean nothing in numerical terms, but the architect might be able to show the designer some other scene in which part of the scene 'stands out boldly' to the degree which he has in mind for the architecture, or where some other part is 'unobtrusive' in the way which he wishes for the auditorium.

The procedure would then be for the designer to measure the luminance of the relative parts of the scene, and also to measure the general adaptation level and thence, by studying the appropriate set of apparent brightness scales, determine the apparent brightness of the 'dominant' platform and the 'unobtrusive' auditorium. For example, suppose the architect shows the executant a scene such as Fig. 8.8, in which he considers the areas A and B, to have the necessary attributes of dominance and unobtrusiveness. First the adaptation of the scene is measured, taken as the luminance, expressed in ft–L, which is the numerical equivalent of the illumination, expressed in lm/ft^2 at the observer's eye. Alternatively, several readings of luminance could be taken with a restricted-angle luminance meter such as a photo-

Fig. 8.8 *Auditorium of Queen Elizabeth Hall, London (Courtesy of Greater London Council. Architect: Herbert Bennett)*

graphic exposure meter, and the adaptation level could be taken as the average of the value so obtained. The individual luminances of the significant areas A and B would then be measured.

Suppose, for example, the adaptation level has been found to be 10 ft–L, and the luminances of A and B are respectively 200 ft–L and 1 ft–L. Reference (in Fig. 8.5) to the appropriate scale of apparent brightness for adaptation level 10 ft–L can then be made. This will reveal that a luminance of 200 ft–L corresponds to an apparent brightness of 350, and a luminance of 1 ft–L to an apparent brightness of 10. These values, 350 and 10, are respectively the necessary apparent brightnesses for the concert platform and for the auditorium walls, and therefore give the prescription, in terms of apparent brightness, to which the designer can then work.

Hewitt, Kay, Longmore and Rowlands (1967) have given a detailed working example of the use of the apparent brightness concept to interior lighting design. Their method, which is more thorough than the simplified examples given above, can be taken as the definitive approach to apparent brightness design at the moment. It is less complex than Waldram's method, but nevertheless achieves a fully adequate degree of accuracy for practical design purposes.

Attention and Distraction

It has been long recognised that human beings turn the eye naturally to light. The biologist uses the term 'phototropism' for the reflex action which all light-sensitive organisms possess. It has, however, proved difficult to demonstrate this effect quantitatively by experiment in relation to human response to interior lighting. Various techniques have been devised by physiologists and experimental psychologists to study the pattern of eye movements. These methods have been applied, for example, to the study of car drivers driving down a street. Most of these techniques require some form of attachment to the head and the placing of electrodes around the eyes, or alternatively the presence of photographic or electronic camera equipment to record eye movements. It is obvious that once the subject knows that he is being observed in this way, and that his eye movements are the subject of study, it is most unlikely that he will behave in a normal and unaffected manner, unless other demands on his attention, as in driving a car, quickly make him forget his part in an experiment. Many investigators have used these highly sophisticated electro-ocular techniques, but it is evident that their use is limited to situations where the awareness of participating in an eye-movement experiment will not adversely affect the results.

A hidden camera, or a hidden observer, can, on the other hand, make satisfactory records, provided that these records can be analysed in an acceptable manner. In an investigation at the Building Research Station, Hopkinson and Longmore (1959) used a technique in which a camera hidden from the observer recorded on moving film the eye movements of the observer confronted with a visual field of which the luminance pattern could be altered as required (Fig. 8.9).

The observer sat on a chair at a distance of ten feet from a large white

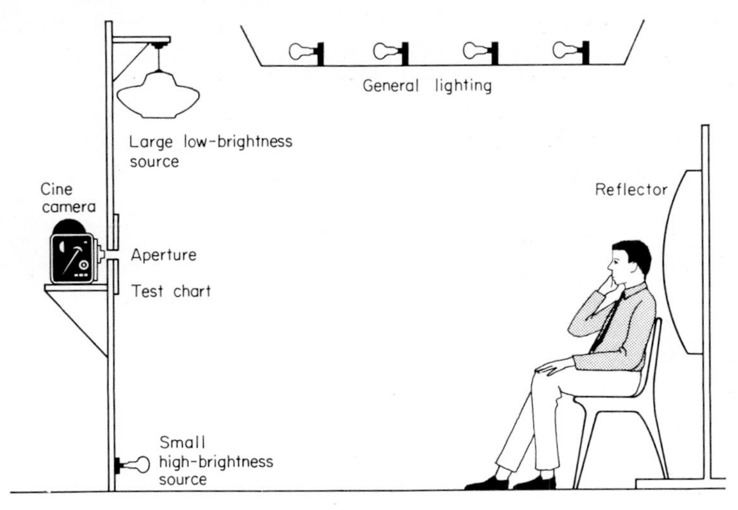

Fig. 8.9 *Arrangement for study of phototropic effect*

screen in the middle of which was placed a concealed aperture through which the recording camera operated. The observer was deceived into thinking that he was required to undertake a certain visual task and he was given no clue whatsoever that his eye movements were being observed. (Such deception, perhaps unethical, is essential for the complete success of the experiment.) Associated in the visual field were various light sources of different size and luminance, which could be switched on or off at will, and in the centre of the visual field was a Landolt ring chart which, the observer was told, constituted the visual task. The subjects were told that the purpose of the experiment was concerned with their accuracy of reading the ring chart. In fact, it was nothing of the kind, and this was merely there to add to the deception. The observer was asked to undertake the reading of this Landolt ring chart and was told that various light sources would be introduced into the visual field. He was told that the purpose of these was to see whether it affected in any way his ability to read the chart. He was asked to treat the test chart as an ordinary reading task, performing it as quickly and accurately as he could and then, having completed the task, to relax until the next visual task was presented. The subject was then left alone and received his instructions from the experimenter in the next room. His eye movements were photographed on the cinematograph camera both while he was reading the chart and also while he was relaxing.

The analysis of the results proved to be much more complicated even than had been expected. The behaviour of the observers was by no means identical. When left alone and unaware of the camera operating, observers very often behaved unexpectedly and there were certain personality factors which occasionally dominated the behaviour pattern and masked the phototropic effects which were being sought. For example, it was expected that when observers relaxed from doing the visual task, their gaze would be directed towards one or other of the bright light sources in the field of view and in most cases this was what happened. On the other hand, in one or two cases observers squirmed and fidgeted in various complex ways, looked at the floor, screwed up their eyes, lent their head right back and looked at the ceiling; and in fact the behaviour patterns in general linked up in rather an interesting way with personality factors revealed in other experiments, particularly those by Witkin (1950) in his studies of spatial orientation.

Those observers who gave recognisable phototropic responses followed a pattern which could eventually be recognised with certainty after sufficient film material had been collected. The observations are summarised in Fig. 8.10, and showed that:

1. A light source in the field of view tended to distract attention from the visual task.
2. On the other hand, if the work itself was bright, attention remained upon it.

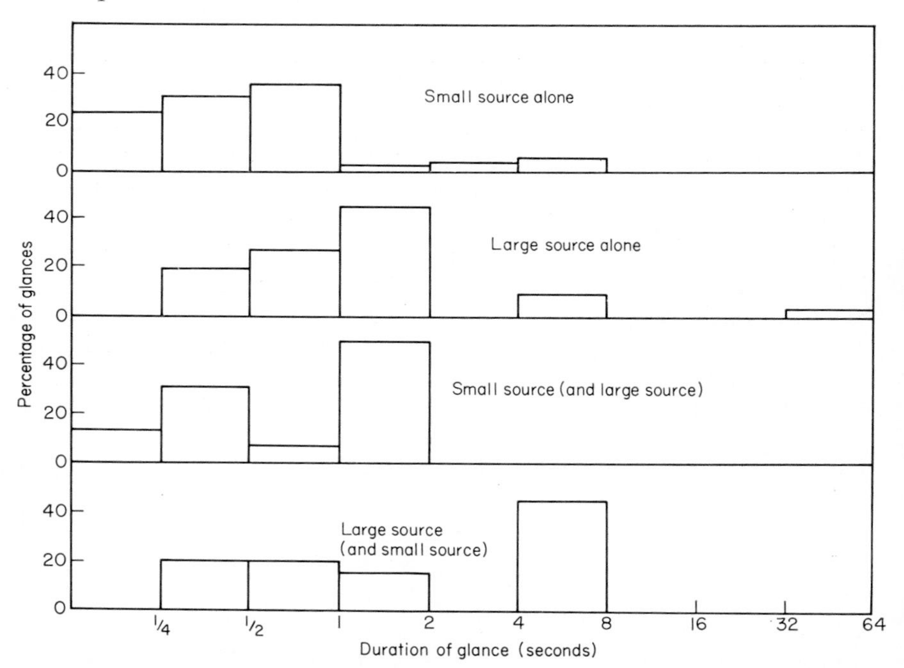

Fig. 8.10 *Analysis of phototropic effect of sources of large area and low brightness, and of small area and high brightness, presented separately and simultaneously in field of view*

Median	0·50 s	1·125 s	0·90 s	1·25 s
Observer occasions	42	23	14	14
Number of glances	59	49	16	20
Total time	33 min 24 s	22 min	13 min 31 s	13 min 31 s
Percentage of time looked at source	2·32%	7·56%	1·57%	6·32%

3. A large source of low luminance tended to hold the gaze for a longer period than a small source of high luminance.
4. The eye movements in the presence of a large source of low luminance consisted of a series of relatively slow traverses towards the source where the gaze was held for many seconds at a time.
5. In the presence of a small source of high luminance, the eye movements, on the other hand, consisted of a series of quick movements towards the source and away again. Sometimes the eyes moved only part of the way towards the source and back to the visual task. Presumably this eye movement pattern was associated with the glare discomfort which resulted from the source of high luminance.

While this pattern of eye movements seems to be characteristic of observers who make any kind of systematic response to different luminance patterns in the visual field, the investigation did not reveal sufficiently precisely at what stage the changeover occurs between emphasis and distraction. The practical outcome however, is, to reveal that attention can be held on work by preferential lighting provided that this preferential lighting is not so excessive that direct discomfort results. Such excessive preferential lighting defeats its own object, because although the phototropic effect is very marked, and the observer will continuously be looking in the direction of the preferential lighting, neverthless the discomfort will have an inhibiting effect. This must necessarily be fatiguing and distressing, even though it has not yet been proved so by direct experiment.

Knowledge of the phototropic effects of preferential lighting can also be applied in the mitigation of unwanted reflections and reflected glare. A specular or semispecular reflection in a glossy table top or a machine will be a cause of phototropic distraction which can be disturbing or even dangerous. If the reflection is small and very bright, the eye movement pattern will be jerky towards and away from the distracting reflection. If the reflection is larger and of lower luminance, the attention will be held where it is not wanted. If, however, such a reflection of low luminance can appear at a point on the machine where attention needs to be held, while not being of sufficient brightness to cause any form of disability or discomfort, it can then be an advantage because it will provide a natural focus of attention to the eyes to draw them to where they should be looking.

There are comparable effects arising from colour and from contrast in the visual field. These have not been sufficiently investigated, although there is some empirical knowledge of their magnitude. In an interior in which the general colour decoration scheme is, for example, of high reflectance but low colour saturation (chroma) as is very common in modern daylight design, a small area of high chroma will attract the attention. This fact is often very useful in designing the colour schemes in areas where people need some guidance because they are not in the environment sufficiently long to be able to learn their way about easily. Chromatropic effects have been used to advantage in hospitals, for example, where patients rarely stay long enough to know their way about instinctively. The intelligent use of high chroma colour on doors to toilets and other places where patients have to learn quickly to go has precisely the same visual effect as the use of preferential lighting.

References

Craik, K. J. W. (1938): 'The Effect of Adaptation on Differential Brightness Discrimination.' *J. Physiol.*, **92**, 406.

Craik, K. J. W. (1940): 'The Effect of Adaptation on Subjective Brightness'. (*Proc. Roy. Soc. B.*, **128**, 232.

Hewitt, H., J. Kay, J. Longmore and E. Rowlands (1967): 'Designing for Quality in Lighting.' *Trans. Illum. Engng. Soc.* (*London*), **32**, 63–89.

Hopkinson, R. G. (1936): 'The Photographic Representation of Street Lighting Installations.' *Phot. J.*, **76**, 323.

Hopkinson, R. G. (1939): 'Discussion on "The Response of the Eye to Light in Relation to the Measurement of Subjective Brightness and Contrast" by W. D. Wright.' *Trans. Illum. Engng. Soc.* (*London*), **4**, 13.

Hopkinson, R. G., J. M. Waldram and W. R. Stevens (1941): 'Brightness and Contrast in Illuminating Engineering.' *Trans. Illum. Engng. Soc.* (*London*), **6**, 37–47.

Hopkinson, R. G. (1948): Proc. Conf. on Lighting and Colour. Council of Industrial Design.

Hopkinson, R. G. (1951): 'Lighting Research at the Building Research Station.' *Light and Lighting*, **44**, 10–20.

Hopkinson, R. G. (1957A): 'Assessment of Brightness—What we See.' *Illum. Engng.* (*New York*), **52**, 211–222.

Hopkinson, R. G. (1957B): 'Subjective Judgment of Brightness.' Proc. 15th Internat. Congress Psychol., Brussels, 103.

Hopkinson, R. G. (1959): 'Adaptation and Scales of Brightness.' Proc. CIE 14th Session (Brussels, 1959), P. 59.19. Commission Internationale de l'Éclairage, Paris.

Hopkinson, R. G. and R. C. Bradley (1959): 'The Estimation of Magnitude of Glare Sensation.' *Illum. Engng.* (*New York*), **54**, 500–504.

Hopkinson, R. G. and J. Longmore (1959): 'Attention and Distraction in the Lighting of Workplaces.' *Ergonomics*, **2**, 321–333.

Hopkinson, R. G. (1960): 'An Experiment on the Assessment of Brightness under "Free Choice" Conditions by a Group of Observers.' *Ergonomics*, **3**, 44–50.

Hopkinson, R. G. (1965): 'A Proposed Luminance Basis for a Lighting Code.' *Trans. Illum. Engng. Soc.* (*London*), **30**, 63–88.

Jay, P. (1967): 'Scales of Luminance and Apparent Brightness.' *Light and Lighting*, **60**, 42–45.

Jay, P. (1968): 'Inter-relationship of the Design Criteria for Lighting Installations'. *Trans. Illum. Engng. Soc.* (*London*), **33**, 47–71.

Lynes, J. A. (1968): 'Beyond the 1968 Code.' *Light and Lighting*, **61**, 101–102.

Marsden, A. M. (1969). Ph.D. Thesis, University of Nottingham.

Pitt, F. H. G. (1939): 'The Effect of Adaptation and Contrast on Apparent Brightness.' *Proc. Phys. Soc.* (*London*), **51**, 810.

Stevens, S. S. (1955): 'The Measurement of Loudness.' *J. Acoust. Soc. Amer.*, **27**, 815–829.

Stevens, S. S. (1956): 'The Direct Estimation of Sensory Magnitudes—Loudness.' *Amer. J. Psychol.*, **69**, 1–25.

Waldram, J. M. (1954): 'Studies in Interior Lighting.' *Trans. Illum. Engng. Soc.* (*London*), **19**, 95–133.

Witkin, H. A. (1950): 'Orientation to Visual and Postural Vertical.' Symposium on Psychophysiological Factors in Spatial Orientation, at Pensacola, Florida, 18–29. US Office of Naval Research, N.A.VEXOS P–966

PART III. APPLICATIONS

9

Codes of Artificial Lighting Practice

Most industrial countries of the world now have codes of recommended artificial lighting practice, which are either issued by an official body or which come from a professional or industrial organisation sufficiently independent of commercial pressures to command respect and authority. The purpose of these various lighting codes has been restricted, until quite recently, to stating in simple tabular form the amount of light (illumination in lumens per unit area) which is considered to be the minimum, or the desirable, level of illumination necessary for proper performance of a wide range of practical visual tasks. Codes of practice usually attempt to give values of illumination level which are likely to be accorded general recognition and which therefore can be regarded as authoritative and capable of being appealed to by either side in a difference of opinion.

Great Britain and the United States of America were among the earliest to issue codes of lighting practice. In both countries these codes were prepared and published by bodies (the Illuminating Engineering Society of Great Britain, and the Illuminating Engineering Society of America) which could speak authoritatively not only for the lighting industry, but also for the user of lighting. The general levels of illumination recommended by these bodies have crept up steadily over the years. This has always been a cause for criticism, and it was undoubtedly one of the factors which led the Illuminating Engineering Society of Great Britain in the early 1930s to set up a committee under A. W. Beuttell, to examine whether a strict analytical and scientific basis for lighting code was possible in order to divorce the situation from any suggestion of commercial pressure. It was as a result of the Beuttell Committee's deliberations that proposals for a lighting code based in terms of the critical contrast and critical detail of a visual task (see Chapter 3) were first put forward by Beuttell himself, and later elaborated in detail experimentally by H. C. Weston. Twenty years later the Illuminating Engineering Society in the United States developed an independent system for codifying its recommended lighting levels based upon the contrast threshold work of H. R. Blackwell, which has also been discussed in Chapter 3. The code of lighting practice in Russia, which unlike the codes in Great Britain and the United

States has statutory significance, is also based on an analytical approach deriving from the work of Meshkov, Shaikevich (1959) and others. Apart from these three attempts to specify illumination levels on a strictly analytical basis, all the other codes of practice in use throughout the world are either, like the British code, derived from the basic work of Weston, or are based upon existing practice and experience.

A comparison of the various codes of practice in use throughout the world, in terms of recommended levels of illumination, immediately invites comment on the differences in the illumination levels recommended for identical tasks. Such differences inevitably give rise to speculation as to the origin of such wide discrepancies.

It has to be remembered that there are at least two distinct ways in which recommended levels of illumination can be specified for any given visual task. First, there is the recommendation based on the absolute minimum illumination in which the visual task can be performed adequately. The Russian code of practice is based upon obligatory minima, and certain government standards in Great Britain (for example, the Factories Act, and the Department of Education and Science Building Regulations for School Teaching Areas) follow this pattern of laying down a minimum level of illumination below which penalties will be incurred for infringement. The specification of obligatory minima has the major disadvantage that when, as is usually the case, design is closely controlled by economies, the obligatory minimum tends to be the design figure and it is too low for this purpose.

The alternative form of specification is a 'recommended' value, well above the obligatory minimum, which is known either by analysis or by experience to yield a level of visual performance above that at which errors or difficulties associated with the lighting can occur. The British IES lighting code, for example, based its recommendations broadly on Weston's experimental work in order to achieve a level of visual performance greater than 90% on Weston's basis, but in the most recent editions of the British IES code (1961, 1968) a direct link with Weston's experimental work is no longer claimed, Instead it is made clear that while Weston's work supplies the fundamental background, in that the relative values of recommended illumination are broadly linked to the visual difficulty of the different tasks in terms of the Beuttell–Weston concept of critical size and critical contrast, the actual values of the levels of illumination now recommended are linked more with current practice of general fluorescent lighting than with any precise recommendations derived from Weston's detailed experimental results. An example of the illumination levels recommended in the 1968 edition of the IES code is given in Table 9.1*

Comparisons between different codes issued in different countries must

* The Illuminating Engineering Society decided that a Code appearing in 1968 and intended to be in current use for five years or more should use SI metric units. The illumination values are therefore in lux and have been rounded off to give a logical series of steps (see page 203).

Table 9.1

IES CODE RECOMMENDATIONS, 1968

GENERAL BUILDING AREAS

The values apply to lighting of the areas in all buildings except where special requirements are indicated in the Schedule

		Minimum service value of illumination (*lux*)	*Limiting glare index*
Entrances	Entrance halls and lobbies, waiting rooms	200	
	Enquiry desks	400	
	Entrance gates and control	100	
	Gatehouses	200	
Circulation Areas	Corridors, passageways	100	22
	Lifts	200	
	Lift lobbies	200	22
	Stairs and escalators	100	13
Staff Restaurants	Canteens, cafeteria, dining rooms		
	General	200	22
	Counters	400	
Kitchens	General	200	22
	Food preparation, cooking, washing-up	400	
	Food stores	200	25
Staff Rooms	Changing rooms, locker rooms	100	
	Cleaners' rooms	100	
	Cloakrooms, lavatories	100	
	Rest rooms	100	
Medical and First Aid Centres	Consulting rooms, first-aid cubicles	400	
	Rest rooms	50	
	Treatment rooms	400	
	Medical stores	200	25
Telephone Exchanges	Manual exchange rooms	200	16
	Main distribution frame rooms	200	25
	Teleprinter rooms	400	19
Outdoor	Car parks	10	
	Main entrances and exits	20	
	Internal factory roads	In accordance with BSCP 1004	
	Stores, stockyards	20	
	Covered ways	50	

also take into account the fact that Weston's experimental data can be interpreted in a number of different ways, and that even if interpreted in the same way, a code of practice can be based on an attempt to achieve 80%, 90%, 95%, 98% or indeed any chosen level of visual performance as defined in terms of Weston's experiments. It is clear, however, that the codes issued by different countries, even though they may be claimed to rest upon the foundation of Weston's work, nevertheless are related to the state of the country's economy, and that the recommended levels of illumination are governed at least as much by what people in the country are prepared to afford for their lighting as by the indication of a strict interpretation of the Weston experimental data. In general, most countries (e.g. Australia) which follow the British IES Code recommend levels of illumination which are closely similar to those adopted in Great Britain. In Russia, although the experimental basis is derived from independent Russian researchers, the recommended levels of illumination, while generally lower than those recommended in Great Britain, stand closely in relation to one another for the different visual tasks. They are lower because the values are laid down as obligatory minima.

The most difficult situation arises in relation to the recommendations made by the Illuminating Engineering Society in the United States of America. The USA recommendations are based on the work of Blackwell which, as has been explained in Chapter 3, demands the translation of contrast threshold data for targets of different sizes into recommendations of illumination level for practical visual tasks. Blackwell has introduced the concept of the Field Factor, a figure used to multiply the illumination level which, under the controlled laboratory conditions of his contrast threshold experiments, results in a given probability of recognition, in order to specify the required visual performance on a practical seeing task. Blackwell's work in its three stages, the contrast threshold experiments, the Field Factor concept, and the Visual Task Evaluator, has been described in Chapter 3. The levels of illumination which have been put forward as recommended practice by the IES of the United States are nearly all very much higher, sometimes by a factor of 3, than those recommended in Great Britain and in other parts of the world. This inevitably has given rise to serious concern and to criticism of Blackwell's work.

Largely as a result of this criticism, Fry (1962) made a detailed analysis of the data obtained by Weston using his Landolt ring cancellation test. Fry shows that if Weston's data are plotted in terms of contrast against background luminance, the curve so obtained for different levels of performance as obtained by Weston lies broadly parallel to the threshold contrast curves. This is in accord with Blackwell's claim for parallelism of threshold and suprathreshold curves. Fry contends that the agreement between the data of the two investigators is extraordinarily good, particularly in view of the difference in their experimental conditions.

Fry's analysis was painstaking and it is unlikely that any other worker has had the inclination or made the time to go through Fry's procedure; it is wise to assume that Fry is right in his statement that the experimental basis of both Weston and the Blackwell data is essentially the same. If this is accepted, then the difference between the American and the British (and other continental countries) recommendations is essentially one of interpretation rather than of foundation material. This was agreed at the session of the Commission Internationale de l'Éclairage in Washington in 1967, when Blackwell's proposed data were adopted for the derivation of a basic relative threshold contrast curve.

Fry supports the principle of determining illumination level requirements by the process of the reduction of the task to threshold visibility by optical means in order to find the standard test object to which it is equivalent. This is one factor in Blackwell's procedure which has been under criticism, but it has now (CIE 1968) been agreed as a basis for international study. If this 'threshold reduction' process is accepted, the next stage is selecting the level of background luminance according to the required standard of suprathreshold visual performance. It is important to note that Fry is content with drawing the scientific conclusions from the data of Blackwell and Weston, and makes no attempt himself to justify the choice of any given particular standard of performance in drawing up a code of lighting levels. In fact, he warns against the attempt to set too high a level of illumination, since this may cause (a) the possibility of direct visual discomfort through too high a task luminance, (b) the possibility that the visibility of some contrasts may actually be reduced at very high levels of luminance, and (c) the obvious fault of recommending levels of illumination beyond those which are economically acceptable.

Fry's analysis of the situation would therefore suggest that the very large differences between the recommended levels of illumination in the American code on the one hand, and the codes of other countries on the other, are due entirely to the question of interpretation in terms of current engineering and commercial practice in the various countries and not to the fundamental experimental work upon which the codes claim to be based.

Weston (1961) also made a study of the basis for establishing codes of illumination requirements, reviewing both the American work and also the basis of the Russian code. His conclusions were briefly as follows. He examined five criteria which may be used in specifying illuminating levels. These are:

1. Adequacy of light for preventing occupational eye strain, unnatural deterioration of sight, and the risk of accidental injury due to bad visibility.
2. Adequacy of light for maximum visual capacity.
3. Adequacy of light for creating an agreeable luminous environment.
4. Adequacy of light for a standard degree of suprathreshold visibility and an associated standard of visual capacity.

5. Adequacy for *different* 'satisfactory' levels of visual performance, each standard being applicable to a particular range in a gamut of visual tasks.

Of these criteria, adequacy for safety etc., is the one which was originally used to prescribe legally enforceable minimum levels of illumination in factories and other places where there is a risk of injury or accident (subject to compensation) being caused by inadequate lighting. Current practice has, of course, advanced far beyond this stage during the past 50 years, and recommendations based on this bare minimum of light have no longer any relation to good current lighting practice.

The use of the second criterion, adequacy for maximum visual capacity, would lead to recommendations of very high levels of lighting indeed if any attempt were made to find at what illumination the most difficult visual tasks could be most efficiently carried out. If, in fact, practice was based on such a criterion, lighting would all have to be designed to this very high figure. This criterion is therefore at the moment not economically attainable.

The third criterion, adequacy for creating an acceptable environment, (sometimes called 'amenity lighting') has now largely superseded the first criterion in the determination of minimum illumination levels legally enforceable in factory working areas, and has led to the recommendation of certain 'amenity levels' in the current editions of the British IES code for areas where there may be no need for critical seeing, and for which there are no 'performance' demands for other than very low levels of illumination. In such areas it is considered that even though visual requirements are very easy to satisfy, a reasonable level of illumination is necessary, well above the bare safety minimum, in order to avoid creating a gloomy and depressing interior. The amenity level is not intended to apply to areas where critical tasks are to be performed, except where it may act as a level of general environmental lighting to be supplemented by local lighting on the task itself to a level to be determined by one of the other criteria.

The fourth criterion, adequacy for a certain standard of visibility and visual capacity, is effectively the scientific basis for the American code. The main practical drawback in choosing this particular criterion is that it can lead to a very wide range of recommended illumination levels in the same building if any conscientious attempt is made to ensure equal levels of visibility or standard of visual capacity for all tasks. A strict interpretation of the Blackwell–American IES criterion is therefore not a practical proposition in a code for interior illumination values. The tendency therefore is to ignore such a code where the illumination values it recommends are either too low for commonsense or too high for economy. A code which invites disobedience is clearly held in only limited respect.

The fifth criterion, a variable standard of performance, has been accepted by the British IES as the most practical way of choosing a range of illumination levels in relation to different visual tasks which are likely to be achieved in good interior lighting. Such a code is more likely to command respect, and this has been proved in the event. A far greater proportion of interior lighting in Britain, and in countries which associate themselves in their recommendations with the British approach, is found to conform to the current code of practice than is the case in the United States. While a certain limited number of 'prestige' lighting installations in America conform to the current American code, by far the greater proportion are more in line with levels recommended and actually achieved in Great Britain and the rest of the industrial world.

Although the Russian code of practice has a somewhat different basis, in that it is drawn up in terms of statutory minima, nevertheless the fundamental basis of the Russian code is essentially the same as that of the British code. This has been discussed by Shaikevich (1958), who suggests that compensation for differences in size of critical detail to be perceived might be achieved by limiting the value of contrast for different size classification of tasks. Even this, however, he admits, would not entirely equalise the visibility of all grades of tasks, nor lead to equal rates of task performance. Standards based on a realistic and variable standard of performance are therefore accepted.

The current code of recommended lighting practice published by the Illuminating Engineering Society of Great Britain (1968) contains recommendations for levels of illumination for over 350 different visual tasks. These values are related to many different varieties of visual activity, not entirely restricted to building interiors, since factory yards, outdoor areas for generating stations, etc., are included, but broadly the recommendations are concerned with the interior environment. The specific values given are restricted to a definite series in which the steps are related to corresponding steps in visual difficulty between one class and the next. Weston's work demonstrated that, in order to achieve a worthwhile improvement in visual performance, steps in the 'illumination series' should be related approximately in a ratio of 1·5 to 1. The recommended steps therefore go in a series 20, 50, 100, 200, 400, 600, 900, 1300, 2000 and 3000 lux, with a basic level of 200 lux for the amenity level, which represents the lowest level recommended for normal interiors. Since a very wide range of visual tasks is listed, it is almost always possible to find directly from the table the level of illumination recommended for any particular industrial and nonindustrial task. Should a particular task not be listed, reference can be made to the classification table in the Code (see Table 9.2).

The Illuminating Engineering Society issues the most complete code of recommended lighting practice in Great Britain at the moment. The British Standards Institution is, however, the official body responsible for the issue of codes of practice of all kinds in Great Britain, and it is likely that in due course the BSI will be responsible, with the cooperation of the IES, for future codes of practice. A British Standard code of practice of electric lighting does, in fact, exist (BSI 1948), and this superseded the code of practice of artificial lighting issued in 1945. This British Standard, however, applies only to dwellings and schools, and the recommendations on illumination

Table 9.2

RECOMMENDED MINIMUM SERVICE VALUES OF ILLUMINATION FOR DIFFERENT CLASSES OF VISUAL TASK

Class of visual task	*Examples*	*Minimum service value of illumination (lux)*
Casual seeing	Locker rooms	100
Rough tasks with large detail	Heavy machinery assembly; stores	200
Ordinary tasks with medium size detail	Wood machining; general offices; general assembly	400
Fairly severe tasks with small detail	Food can inspection; clothing, cutting and sewing; business machines; drawing offices	600
Severe prolonged tasks with small detail	Fine assembly and machining; hand tailoring; weaving silk or synthetic fibres	900
Very severe prolonged tasks with very small detail	Hosiery mending; gauging very small parts; gem cutting	1300–2000
Exceptionally severe tasks with minute detail	Watchmaking; inspection of very small instruments	2000–3000

levels are not put forward in any great detail. In practice, therefore, the British Standard is at the moment not in wide use, whereas the British IES Code represents current lighting practice.

Codes of Recommended Practice for Visual Comfort

The British Illuminating Engineering Society has gradually elaborated its code of recommended lighting practice with succeeding editions so that the most recent editions (1961, 1968) include detailed sections on the wider problems of design for good lighting over and above the simple recommendation of necessary levels of illumination. Above all, the IES has recognised that good lighting under current conditions of moderate to high illumination levels depends not only upon the light on the work, but also upon the distribution of light in the whole visual field. The reason for the critical importance of the whole field began to be recognised when fluorescent lighting was first employed on a large scale. It had never been possible with filament lighting to install general lighting throughout large offices and workshops to a level much above 100 or 150 lux (or 10 or 15 lm/ft^2). When fluorescent lighting, with its advantage in luminous efficiency of approximately 3 to 1, began to be commercially available, lighting installations were immediately designed to levels of the order of 500 lux (or 50 lm/ft^2). Instead of giving immediate satisfaction as had been predicted, these installations gave rise to widespread complaint, the reason for which was not immediately detected, because the complaints were expressed vaguely, often in terms of 'too much light'.

As a consequence, current lighting codes, but particularly those of Great Britain and of Australia, supplement their recommendations of illumination levels with detailed instructions on the elimination of direct glare from lighting fittings. The Australian lighting code goes one stage further, in making recommendations for the elimination of glare reflected from the working surface.

The British IES recommendations for the control of glare are based on the Glare Index system (see Chapter 4). The IES code gives a table of recommended limiting values of Glare Index related to the character of the environment (rather than specifically to the visual tasks in the environments) and these values are tabulated alongside the recommended levels of working illumination. The way in which they were established is described in more detail in Chapter 4. Table 9.3 gives a selected list of recommended limiting values of Glare Index to indicate the way in which the Glare Index is related to the type of environment.

Table 9.3

RECOMMENDED LIMITING VALUES OF GLARE INDEX (IES)

Type of Environment	*Limiting Glare Index*
Store rooms	28
Garages and parking areas	28
Farm buildings	25
Assembly shops (average conditions)	25
Assembly shops (precision instruments)	19
Laboratories	19
Printing works (composing room)	19
Libraries	19
General Offices	19
Drawing offices	16
School classrooms	16
Hospital wards	13
Hospital operating theatres	10

It will be seen that the principle is broadly that the more important it is to avoid distraction of the operative, or the less opportunity the occupant has of directing his gaze away from an offending light source, the lower the recommended limiting value of Glare Index; while a higher value is recommended in situations where people are not likely to be much concerned about the refinements of their environment.

The British IES system is comprehensive in the sense that it can be used to compute the Glare Index for a very wide range of situations in which the lighting fittings are arranged in conventional symmetrical arrays. Because of the attempt to make it universal, it is a more complicated system than is strictly necessary for the great majority of conventional lighting installations. The system was devised specifically in order to permit the use of any intelligently designed lighting unit, even one of high luminance, provided such fittings were used in such a way that their cumulative effect did not exceed the recommended Glare Index limits. Previous methods of glare limitation based on limiting the luminance of the lighting unit as seen by the observer militated against many designs of a sophisticated nature which experience had shown were acceptable when intelligently used. This is particularly true of units which employ refracting or diffusing prisms, which may have a high individual maximum luminance, but which when used correctly in a well-designed installation can meet the IES Glare Index limits while also introducing a welcome degree of 'sparkle'.

Nevertheless the IES Glare Index system has been criticised for its complexity, and alternative methods claiming to simplify the situation have been proposed. Many years before the IES system was introduced, both the British and the American Illuminating Engineering Societies, and other organisations in other parts of the world, recommended glare control in terms of the limitation of the luminance of the lighting units. Among the regulations and codes may be mentioned the British Regulations on School Building (Ministry of Education, 1959) and the American Illuminating Engineering Society's 'Scissors Curve' (1967). The School Building Regulations give a straightforward limitation of luminance of the lighting unit within certain specified angles, the assumptions inherent in this recommendation being that classrooms will be of certain dimensions with surfaces of high reflectance (related to the daylight design), and for this reason the recommendations can be exceedingly simple. The American Illuminating Engineering Society necessarily has to have a slightly more complex system than this, but even so the recommendation consists simply of two straight lines on the diagram (Fig. 9.1, 9.2).

The control of glare is dealt with in the Australian Code (1965) by giving alternative recommendations for bare-lamp or diffusing fittings and those fittings with opaque screening. The requirements for the first group are given in a table of maximum permissible luminance in relation to room width and length (as functions of mounting height of fitting above eye level), and for the second group in terms of shielding angle requirements for lamps with luminance either below 10 cd/in^2 or greater than 10 cd/in^2 (15 500 cd/m^2). The luminance limits of the bare lamps or diffusing fittings were derived from work by Harrison and Meaker, in U.S.A., and range from $\frac{1}{2}$ to 6 cd/in^2 (775 to 9300 cd/m^2) according to the size of the room for symmetrical fittings or

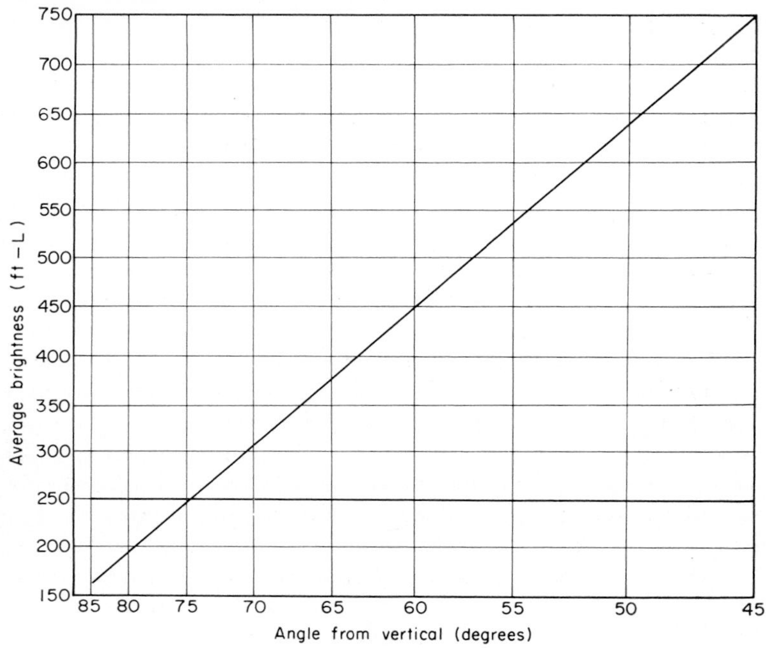

Fig. 9.1 *'Scissors curve' diagram introduced by American IES in 1956 for luminaires for office and school lighting installations* (*From Beggs 1967*)

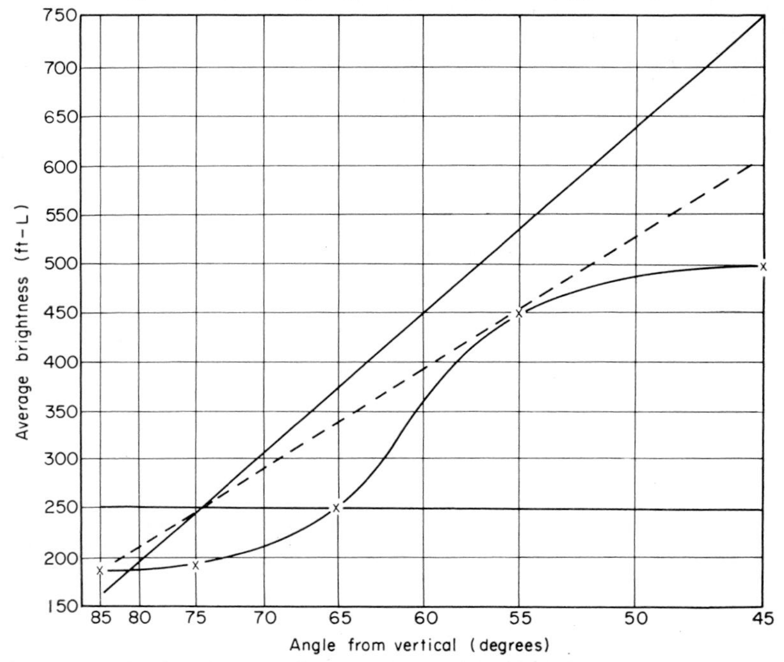

Fig. 9.2 *Example of application of 'scissors curve' diagram to the luminance distribution of a luminaire* (*From Beggs 1967*)

long fittings seen sideways on, while for 'elongated luminaires' seen end on, an increase of luminance up to 2 cd/in^2 (31 000 cd/m^2) to 8 cd/in^2 (124 000 cd/m^2) is permissible.

The recommended shielding angles for opaque fittings are given for four broad classifications of interior or purpose in relation to mounting height and lamp light output. Fittings of this type for a source of luminance over 10 cd/in^2 are not recommended for schools, offices, etc., but for other purposes the minimum shielding angles vary from 20° to 40°, with the proviso that no screening is required for a lamp with a luminance less than 10 cd/in^2 in a storeroom or passage provided that it cannot be seen from a working area.

A further restriction on the use of incandescent lamps and other lamps having a luminance of over 10 cd/in^2 in open fittings is placed by specifying in a further table the minimum mounting height above the floor of such fittings in relation to the lamp light output. The tabulated recommendations assume that the lightnesses of the finishes of the room surface will be such as to achieve the kind of luminance distribution in the interior generally regarded as acceptable (described in a separate section), but no data are given for any variation of surface reflectances, beyond the warning that dark surfaces will require a lower limiting luminance than that given in the table provided. A further warning is given that in rooms where there is substantial horizontal viewing (e.g. classrooms) the luminance of fittings should be at least one step lower than that given in the tables.

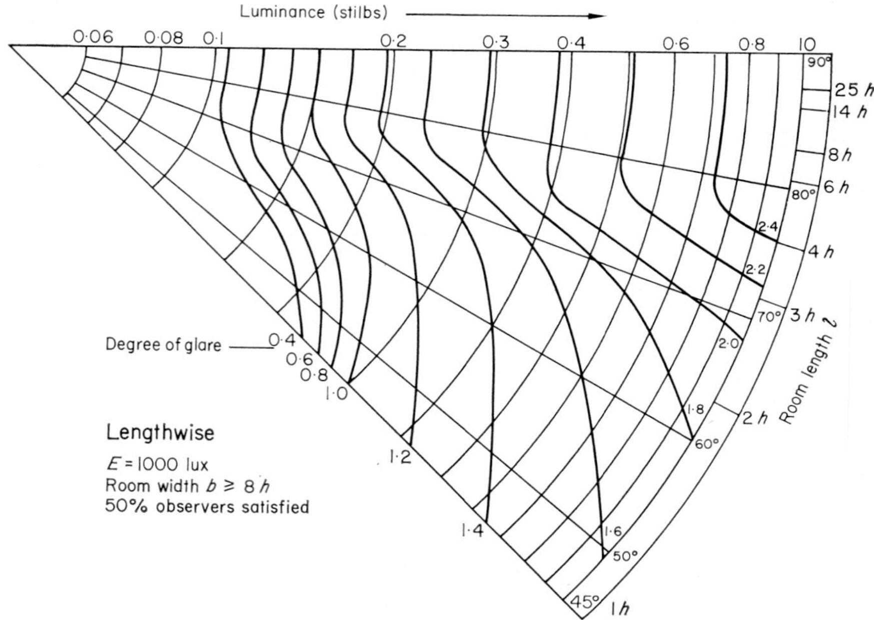

Fig. 9.3 *Relationship between fitting luminance and angle of view for different degrees of glare (lengthwise) (From Bodmann* et al. *1966)*

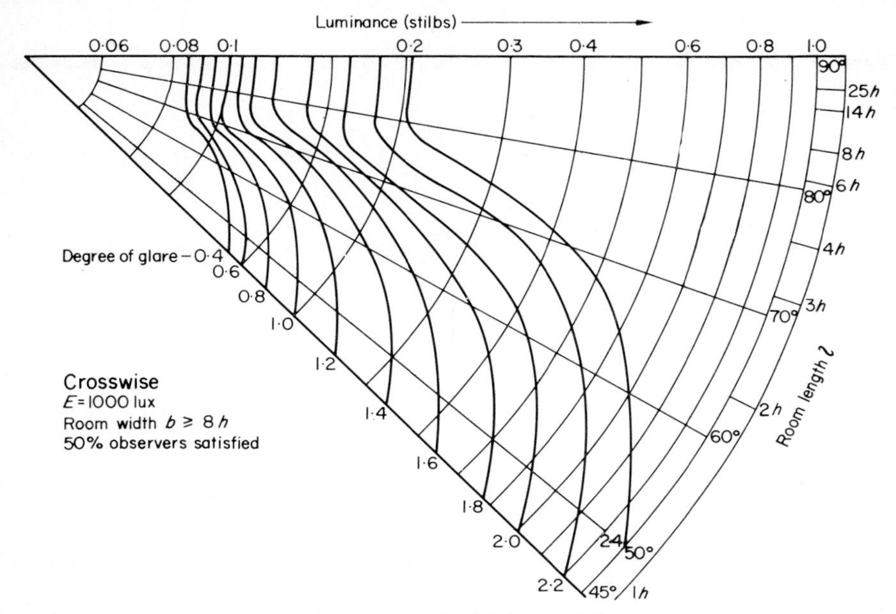

Fig. 9.4 *As Fig. 9.3 (crosswise)*

A system such as the Australian one is very simple to apply provided the manufacturer supplies the necessary luminance data at various angles, but being of necessity one in which the data must be compressed and averaged, it is not possible to use it to exploit the possibilities of refinements in design of the lighting installation or of the interior decorative scheme.

The system advocated by Bodmann and his colleagues (1963 *et seq.*) in Germany is one of luminance limitation, but is slightly more sophisticated than the American 'Scissors Curve'. Bodmann has derived a scale of glare arbitrarily from numbers given to glare criteria, and from observations of a model interior has related fitting luminance at angles between 45° and the horizontal to degrees of glare (see Fig. 9.3). In use, the angle scale is entered at a value represented by room length as a multiple of mounting height, and the maximum luminance at this angle is read off in relation to the degree of glare. Alternatively, the 'glare rating' of a fitting with a given luminance distribution in a room of given length can be read off. Separate diagrams are required for different room widths, for different types of fitting, and for crosswise as against endwise mounting (see Fig. 9.4). A conversion diagram is provided for adjustment of glare rating in relation to illumination level on the working plane (Fig. 9.5) and percentage of observers satisfied (Fig. 9.6).

A Lighting Code Based upon Luminance Considerations

Codes of recommended lighting practice at the present time are based on two main factors: necessary illumination level for adequate visual perform-

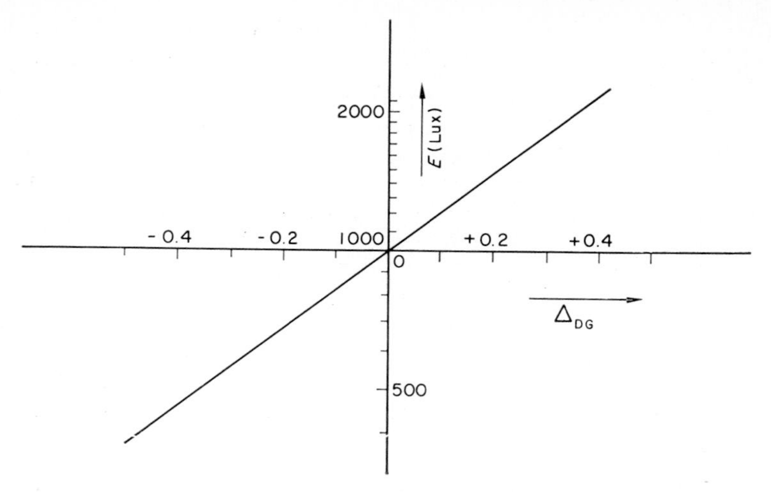

Fig. 9.5 *Shift of the degree of glare* Δ_{DG} *with varying illumination level* (*From Bodmann* et. al *1966*)

ance or for amenity, and freedom from glare discomfort. These two factors are not necessarily dependent and interlinked in the design and specification of a lighting installation. On the other hand, all the research which has been undertaken in recent years demonstrates that good lighting, in terms both of visual performance and of visual comfort, depends upon an integrated concept of lighting design rather than a concept in which illumination level on the one hand and discomfort glare on the other hand are treated by means of independent methodologies.

The need to specify the characteristics of the whole installation has been

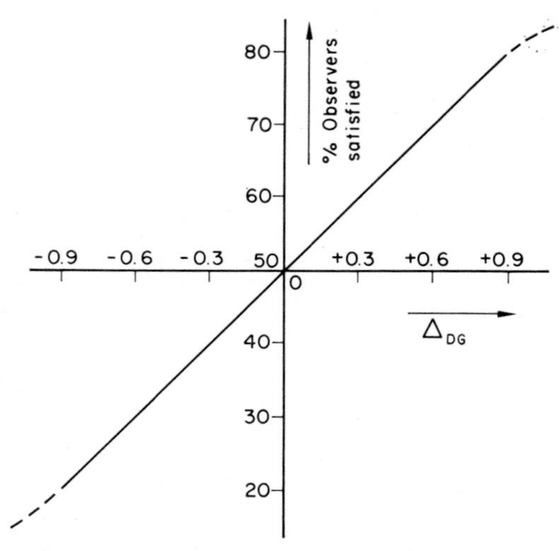

Fig. 9.6 *Shift of the degree of glare* Δ_{DG} *with varying percentage of observers satisfied* (*From Bodmann* et al. *1966*)

recognised for a great many years but the implementation of this recognition in terms of a practical code has so far eluded lighting designers. Hopkinson (1965) proposed a basis for a lighting code in which the luminance distribution of the visual field, as the end-point of good lighting, might be specified in terms of a lighting system incorporating what he called 'building lighting' together with 'work lighting', the first to define the environmental conditions and the second to meet special exigencies of the task itself. Hopkinson's proposals were based upon a long series of experiments at the Building Research Station dating from 1948 (Hopkinson 1948 *et seq.*), the stages of which were as follows:

1. The confirmation that the relation between illumination level and visual performance shows diminishing returns above a certain level, and that therefore above this level the exact value of illumination is relatively unimportant.
2. The demonstration that the concept of apparent brightness as a function of a state of adaptation applies in interior lighting, and that consequently the level of illumination is only one of the factors which govern the overall degree of satisfaction from the lighting.
3. The demonstration that the factors which govern discomfort glare can be quantified on the basis of subjective assessments, and that methods for the control of glare can therefore be devised.
4. The determination of the relation between the light on the task and on the surrounds to the task in relation to visual comfort as well as efficiency.
5. The determination of the upper luminance and luminance ratio limits for visual comfort, especially from large areas.
6. The indication that attention is held by preferential lighting on the object of regard, and that distraction is caused by bright areas away from this object of regard.

During the period that these studies were continuing at the Building Research Station, investigations in other countries, especially the Netherlands and Germany, were being undertaken with essentially similar objectives, and this work (Balder 1957, Bodmann *et al.* 1963 *et seq.*) provided independent confirmation of many of the findings.

Briefly, Hopkinson's proposal for a luminance code is that good lighting can be specified in terms of the absolute values and the ratios of the luminance of the general environment and of the visual task (if any). The task luminance should be decided by the requirements for visual performance as determined by the Beuttell–Weston concept of critical detail and critical contrast, but with the proviso that an upper limit of luminance of the task should not exceed 100 ft–L (350 cd/m^2) except in special circumstances. Guidance can be given to assist in concentrating attention on the work by suggesting that the relative luminance of the task to its immediate surrounds and to the general surroundings of the environment should be of the order of 10:3:1, although there is considerable freedom around these average ratio values. Any extension of this range of values, however, could lead to the possibility of discomfort glare unless exceptional skill and care are used in the lighting design.

In practice, this normally requires the implementation of artificial lighting in the form of some kind of dual system. The environmental lighting (or 'building lighting' as it is described by Hopkinson) is conceived as a uniform or near-uniform system in which the illumination distribution and level, coupled with the reflectance characteristics of the room surfaces, are combined to give a luminance value and distribution which relates to the luminance of the visual task in the ratios specified above. The preferentially high luminance of the work, so required in order to maintain attention on the work and to give the optimum visual comfort conditions, can be obtained either by a special system of preferential local lighting, or, as in many practical circumstances, by the fact that the task will have a preferentially high reflectance (white paper) or, in some cases, by a combination of directional light on the work and diffuse lighting for the environment, built into the same overall system.

Hopkinson's ideas have been worked out in some detail in relation to the lighting of hospital wards (see Chapter 10) where preferential lighting for patients in their beds, supplemented by environmental lighting in the ward itself, to give luminance ratios consistent with Hopkinson's proposals, has been shown to be satisfactory for its purpose, and has consequently been adopted by the Ministry of Health in Great Britain as recommended practice. It is probable that the implementation of Hopkinson's proposals will proceed in this way, that is, by the working out in practical terms of the application of the principles to particular requirements in different kinds of buildings, with the eventual codification of these different practical applications in a comprehensive code of lighting practice.

Meanwhile, these proposals have been studied by the Illuminating Engineering Society, and the principles behind them have been incorporated in the latest edition of the IES Code to the extent that surface reflectances have been recommended which, with a general lighting installation, will create very roughly the luminance patterns proposed by Hopkinson for the task and the environment. It is recommended in the Code that the reflectance of the immediate background to the task in working interiors should be a quarter to a half of the task reflectance, while the reflectance of the roofs and ceilings should be not less than 70%, that of the main surfaces of the walls not less than 40%, and that of the floors between 20% and 40%. The lighting of the room, however, is only specified in terms of illumination level at the work and in limiting Glare Index for the whole environment, as in the 1961 edition. Thus we are still a long way from seeing the main specification of work and environment surfaces purely in terms of luminance, although a suggestion is made in Appendix 3 of the Code (which is intended to give an indication of the current position of research not sufficiently well established to form part of a code of practice) does suggest a range of luminance ratios which could be used as a basis of design.

Since Hopkinson's proposals, Lynes and his colleagues have put forward the results of experiments which show that the effectiveness of lighting is governed by other quantitative parameters than the illumination on a horizontal plane (Lynes *et al.* 1966 *et seq.*) In this work Lynes follows ideas first developed by, among others, Gershun, and more recently followed up in Russia by Epaneshnikov, which demonstrate that the subjective effectiveness of lighting is governed more by the mean spherical illumination at a reference point than by the illumination in the reference plane (normally the horizontal plane). It has been suggested that the 'building lighting' discussed above could be specified in these terms. In addition, Lynes develops the concept of illumination as a vector quantity, having both direction and quantity, and shows how the vector quantity of the illumination in an interior can be used to define the modelling characteristics of the lighting, and how these characteristics can be predicted and made the object of quantitative design. He and his colleagues have also determined the acceptable directions and strengths of modelling in terms of ratio of vector illumination to scalar illumination, and it has also been proposed that future codes should take account of these preferences in recommendations.

These studies of Lynes and his colleagues take the subject of luminance design a stage ahead of Hopkinson's proposals, although they do complicate the problem of welding all the proposals into a workable lighting code. Hopkinson had shown that his work led in practice to a simple form of code, which would consist, as in the application to hospital wards, of simple recommendations for the amount and distribution of the building lighting, for the reflectances of the room surfaces in relation to those of the critical visual tasks, and for selective lighting on the work expressed in terms of quantity and direction. Lynes's ideas would call for a more complicated system of specification. In addition, Lynes's work, and that of Epaneshnikov also, while demonstrating that the subjective estimation of brightness is certainly closely influenced by the spherical illumination rather than the illumination on a horizontal plane, does not take into account the influence of the luminance of the visible sources and of other dominant features of the visual field. Hopkinson (1951) showed that the subjective estimate of the amount of light present in an installation is much greater when the sources are visible than when they are not, and that this estimate is also affected to a high degree by the luminance of the visible sources. Visible filament sources, for example, were found to create an impression of more light than hidden or partly hidden fluorescent sources.

The same value of spherical illumination can result from an interior with visible bright sources as from one with larger sources of low luminance. Consequently, if Hopkinson's results are valid, there is little advantage to be gained from Lynes's proposal to use spherical rather than horizontal illumination as a criterion of lighting quantity.

The situation at the moment is therefore interesting but not at a point close enough to resolution to warrant any hope that a code of lighting practice based on luminance distribution will be developed in the near future. A great deal of work has to be done to establish the validity in practice of the various proposals that have been put forward.

The proposals by Lynes and his co-workers for the use of the vector concept for defining the directional properties of lighting deserve careful study. Their work has led them to propose that preferable modelling of a solid object, such as a face, can be predicted entirely by the ratio of the vector illumination to the scalar illumination, and the direction of the vector. That is, the modelling is independent of whether the directional illumination comes from one source or from two or several sources, provided that the vector direction and quantity is the same. This finding, at first evaluation, is clearly invalid, because the eye is not an integrating device, as the concept demands, but an analysing device, and conditions can easily be demonstrated in which the vector-scalar ratio from two widely spaced sources gives entirely different modelling from that from a single source with the same vector-scalar ratio and directional values. However, provided that the precision claimed is not high, and the range of conditions is limited (for example, the range of source angles over which the vector is evaluated is restricted, and the principle is not applied to sources below a certain solid angle as subtended at the object), the method offers a quantitative procedure for the specification and prediction of the modelling characteristics of lighting, as Jay (1968) has shown. Fischer (1969) is investigating the range of angles and source size over which the vectorscalar concept is valid, using a variety of solid objects for modelling criteria, and when this work is completed, the value of the concept can be better assessed.

References

Australian Standard Code for the Artificial Lighting of Buildings (1965). AS: CA 30–1965, Standards Association of Australia, Sydney.

Balder, J. (1957): 'Erwünschte Leuchtdichten in Büroraümen.' *Lichttechnik*, **9**, 455–461.

Beggs, S. S. (1967): 'Prediction of Discomfort Glare.' *Light and Lighting*, **60**, 245–253.

Bodmann, H. W., G. Söllner and E. Voit (1963): 'Bewertung von Beleuchtungsniveaus bei Verschiedenen Lichtarten', Proc. CIE 15th Session (Vienna, 1963), Vol. C ,502–509. Commission Internationale de l'Eclairage, Paris.

Bodmann, H. W. and G. Söllner (1965). 'Glare Evaluation by Luminance Control.' *Light and Lighting*, **58**, 195–199.

Bodmann, H. W., G. Söllner and E. Senger (1966): 'A Simple Glare Evaluation System.' *Illum. Engng. (New York)*, **61**, 347–352.

Bodmann, H. W. (1967): 'Quality of Interior Lighting Based on Luminance.' *Trans. Illum. Engng. Soc. (London)*. **32**, 22–40.

British Standard Code of Practice CP3 : 1945, Chapter VII F. 'Provision of Artificial Light (Houses, Flats and Schools)'.

CIE (1968): Report of Meeting of Committee E. 1.4.2. (Visual Performance). Proc. CIE 16th Session (Washington, 1967). Commission Internationale de l'Éclairage, Paris.

Fischer, D. (1969). Private communication to the authors.

Fry, G. A. (1962): 'Assessment of Visual Performance.' *Illum. Engng. (New York)*, **57**, 426–437.

Hopkinson, R. G. (1948): 'Brightness Contrast and Glare.' Proc. Conf. on Lighting and Colour. Council of Industrial Design.

Hopkinson, R. G. (1951): 'The Brightness of the Environment and its Influence on Visual Comfort and Efficiency.' Proc. Building Research Congress, Div. 3, Part III, 133–138, HMSO, London.

Hopkinson, R. G. and J. Longmore (1959): 'Attention and Distraction in the Lighting of Work Places.' *Ergonomics*, **2**, 321–334.

Hopkinson, R. G. and R. C. Bradley (1959): 'The Estimation of Magnitude of Glare Sensation.' *Illum. Engng.* (*New York*), **54**, 500–504.

Hopkinson, R. G. and R. C. Bradley (1960): 'A Study of Glare from Very Large Sources.' *Illum. Engng.* (*New York*), **55**, 288–294.

Hopkinson, R. G. (1963). 'Architectural Physics—Lighting.' HMSO, London.

Hopkinson, R. G. (1965): 'A Proposed Luminance Basis for a Lighting Code.' *Trans. Illum. Engng. Soc.* (*London*), **30**, 63–88.

Illuminating Engineering Society, London (1961, 1968.) The IES Code: 'Recommendations for Good Interior Lighting.'

Illuminating Engineering Society, New York (1967): IES Handbook, Chap. 2.

Jay, P. A. (1968). Interrelation of the design criteria for lighting installations. *Trans. Illum. Engng. Soc.* (*London*), **33**, 47–71.

Lynes, J. A., W. Burt, G. K. Jackson and C. Cuttle (1966): 'The Flow of Light into Buildings.' *Trans. Illum. Engng. Soc.* (*London*), **31**, 65–91.

Lynes, J. A., with C. Cuttle; W. B. Valentine and W. Burt (1968): 'Beyond the Working Plane.' Proc. CIE 16th Session (Washington, 1967), Paper 67.12. Commission Internationale de l'Eclairage, Paris.

Ministry of Education (1959). Statutory Instrument 1959 No. 890: 'The Standards for School Premises Regulations.' HMSO, London.

Petherbridge, P. and R. G. Hopkinson (1950): 'Discomfort Glare and The Lighting of Buildings.' *Trans. Illum. Engng. Soc.* (*London*), **15**, 39–79.

Shaikevich, A. S. (1958): 'Classification of Visual Tasks.' *Svetotekhnika*, **12**, (5) 13–20.

Shaikevich, A. S. (1959): 'Extending the Study of Control Standards for Industrial Lighting.' *Svetotekhnika*, **13**, 8–12.

Weston, H. C. (1961): 'Rationally Recommended Illumination Levels.' *Trans. Illum. Engng. Soc.* (*London*), **26**, 1–16.

10

Studies of Some Lighting Design Problems: I. Luminance in Drawing Offices and Hospitals

During recent years it has been shown that better lighting results when the whole pattern of the visual field is studied, rather than the light on the working plane alone. The problem has been to devise methods of design which will permit the prediction of the distribution of luminance in the visual field, and the problem has not so far been solved. It is largely for this reason that the majority of lighting installations are still designed around the illumination level on the working plane.

While basic studies on illumination level, glare and brightness distribution were being undertaken at the Building Research Station, work was being done in parallel to implement the recommendations which were arising from the basic studies. This work was concerned primarily with schools and hospitals, and to a lesser extent with industrial lighting, especially drawing offices.

The implementation of lighting principles always calls for compromise, sometimes to the extent that little remains of the basic ideas. Difficulties of power distribution are often of overriding importance; in the development of lighting based on the concept of environmental 'building lighting' with selective 'local lighting' on the work, it may be impossible to bring power to the work, and so the selective lighting can be achieved, if at all, only by the design of the light distribution from the nearest convenient source of power.

It is important that misunderstanding does not arise on this question of local lighting. It is by no means necessary for local lighting to be provided on any but the most difficult and exacting tasks. The necessary phototropic effects to direct attention on the work often arise quite normally from the fact that people are working on white paper placed on a desk of lower reflectance, so that the general lighting itself provides the necessary distribution of luminance in the field of view for visual comfort and attention.

The Design of Lighting for Drawing Offices

Collins and Langdon (1960) conducted a survey related to the design of drawing offices. One of their findings was that draughtsmen often express the preference for an adjustable local lamp on their work. In the course of this survey, they made photometric measurements in a number of drawing offices of different design and related them to the overall assessments of lighting quality made by the draughtsmen. They reported that the office in which the artificial lighting received the highest ranking for overall satisfaction was one in which the general lighting was of only a moderate level of illumination (about 12 lm/ft^2 or 130 lux average) but in which local lights were provided on each drawing board. The next most satisfactory from the point of view of artificial lighting provided a level of general lighting from two to seven times this amount. The survey also showed that draughtsmen do not spend all their time drawing; on average a draughtsman spends only about 30% of his time actually drawing, while 44% of the time is spent at or near the board in calculation or consultation. (The actual proportions vary somewhat with seniority, but it is significant to note that draughtsmen themselves greatly overestimated the amount of time which they spend at the drawing board.)

The interpretation of the results of this survey as a guide to design is limited. The findings naturally depend upon the characteristics of the installations which were actually studied and do not indicate of themselves, without the confirmation of analytical studies, what the optimum form of design should be.

One interpretation of the results of the survey would be that the optimum form of lighting for a drawing office should be general lighting giving a moderate level of illumination evenly over the whole office, sufficient for auxiliary visual tasks such as reading or operating a slide rule, together with the addition of local lighting at each drawing board under the control of the draughtsman himself. On the other hand, a relatively high rating was given in the survey to drawing offices with general lighting alone (no local lighting) provided this produced a high level of illumination. This could well be interpreted as support for those who advocate a high level of general lighting for this kind of visual task.

The IES Code of Recommended Lighting Practice (1968A) recommends a general level of illumination of 400 lux* throughout the office, with 600 lux actually on the drawing boards, together with a limiting Glare Index of not greater than 16. Lighting to these standards represents a considerable improvement on much of current practice, but it is also likely that this standard can be improved still further. The recommended level of general illumination is probably satisfactory provided the level of preferential lighting is increased beyond the recommended figure of 600 lux.

Since the survey was completed, examples have been seen of general

* See footnote on p. 198.

lighting in drawing offices in which particular attention has been paid not only to providing a high level of illumination (of the order of 600–900 lux) but to the strict elimination of glare discomfort both from the ceiling lighting units and from reflections from the surface of the working area. Such an installation is that in the new Toronto City Hall (Fig. 10.1). The lighting is indirect, but designed in such a way that the soporific effects of a uniform luminous ceiling are avoided, and, by virtue of the moderate value of floor reflectance, a satisfactory luminance pattern is seen by the draughtsman. These examples would suggest that well-designed general lighting may give as great a user-satisfaction as a combination of moderate general lighting plus individually controlled local lighting. Further work remains to be done before any dogmatic statements of the relative merits of the two systems can be made. It is probably true, however, that both systems at their best give complete user-satisfaction.

A word of caution is necessary concerning the use of the limiting Glare Index as a figure of merit for the design of drawing office lighting. Techniques now exist which make use of specular reflecting surfaces either in individual lighting fittings or as part of the ceiling surface to direct light down from the lamp on to the drawing board without diffusing any appreciable quantity towards the eyes of the occupants of the office. Consequently the whole ceiling *including* the lighting devices appears relatively dark while the drawing boards are brilliantly illuminated. The contrast between the dark ceiling and the brightly lit board can in some circumstances be troublesome. The imbalance of brightness in the office causes the boards themselves to be uncomfortably bright areas in a dark and gloomy interior. Yet this system of lighting has a limiting Glare Index on the IES system well inside the recommended limits. This is an example where the Glare Index taken in isolation does not give a complete specification of the visual comfort situation. It does limit direct glare, which is its purpose, but it is necessary also to ensure that there is a correct luminance ratio between ceiling, drawing paper, and general surroundings in the room. Collins and Langdon have suggested that relative values of 7:10:1 for ceiling, drawing board, and general surround will be found satisfactory. It might be advisable to provide an intermediate surround of relative luminance 3 on this scale between board and general surround to optimise the situation. An experimental study at the Building Research Station at the present time is directed to the examination of preferred ratios of ceiling luminance to task and lighting fitting luminances.

Fluorescent lighting in its modern form is entirely satisfactory for the majority of drawing office installations. There may be a small proportion of workers who express discomfort in working in fluorescent lighting due to flicker (see Chapter 5). These people will suffer discomfort in any form of intermittent light source including watching television. There is little that can be done to alleviate their symptoms provided that lamps and control gear are carefully maintained. In a large installation, the operation of lamps on a 3-phase or split-phase system (see Chapter 5) is usually an adequate precau-

Fig. 10.1 *Drawing Office, Toronto City Hall (Courtesy of G. Franklin Dean and the Toronto Hydroelectric System, Ontario, Canada)*

tion against possibility of detectable flicker, provided old and unsatisfactory individual lamps are eliminated.

Experience seems to indicate that little difficulty arises in the use of filament local lights in combination with fluorescent general lighting, contrary to what might be expected. Easily adjustable local lighting with filament lamps is certainly preferable to fluorescent lighting mounted in fixed units placed above the board. The virtue of local lighting, such as it is, is not only in the preferential level of luminance which it provides, but also in the fact that the draughtsman can adjust it to the particular area on which he is working, both as regards quantity and direction.

The Lighting of Hospital Wards

The hospital ward is one interior in which it has been possible to work out a complete lighting specification in accord with the principles of luminance design. The study began in 1948 when Llewelyn-Davies and his colleagues of the Division of Architectural Studies of the Nuffield Foundation conducted an investigation into the function and design of hospitals, which led to new planning of ward layout for efficient nursing and hospital organisation. The Building Research Station cooperated at different stages in this study, and it was from this work that new proposals for the artificial lighting of hospitals were developed. These proposals were subsequently examined by a Joint Committee on Lighting and Vision of the Medical Research Council and the Building Research Board and were adopted by the Ministry of Health*. The practical details of the implementation of these proposals were at a later stage investigated by the Technical Committee of the Illuminating Engineering Society of Great Britain, and guidance on the design of hospital ward lighting is given in a technical report of the IES (1968B).

These recommendations on the artificial lighting of hospitals are based on a combination of experimental work specifically in the hospital context, and proposals deriving from the basic BRS studies. At the outset, it was obvious that there would be a conflict between the lighting system which would be best for the nursing and clinical staff. Among the patients are those who may be quite ill and who wish to sleep or rest for most of the time while the artificial lighting is in operation. These people therefore should not be subjected to high levels of luminance. Instead they need to have a low level of lighting so that areas which they may have to lie and gaze upon for hours on end are well below any discomfort limits. Some patients, on the other hand, are quite well and wish to engage in activities, either on their own (reading, sewing, etc.), or with others.

The nursing and clinical staff may have exacting visual tasks. They need to mark up and read charts and instructions, and of course, they need to be able to see easily to make clinical examinations, dress wounds, etc. They also need to be able to recognise instantly from the appearance and colour of a patient whether an emergency has arisen. Furthermore, all occupants in the ward require a pleasant character and cheerfulness about the lighting during the evening before 'lights out'.

* This Ministry has subsequently been succeeded by the Department of Health and Social Security but recommendations referred to below were all made by the former Ministry.

Quantity of Light

It is clear that the lighting design must compromise between high levels of lighting for some and low levels for others. The compromise was achieved partly by an analytical approach and partly by trial and modification of the systems indicated in this way. The conflict between requirements of staff and requirements of sick patients was resolved in the following way. Hopkinson's (1949) basic studies on illumination level and ease of visual performance showed that a visual task which can be conducted with ease in an illumination of 100 lm/ft^2 (or, say, 1000 lux) can still be conducted, though with difficulty, in an illumination of 1/30th of this value, that is, not less than 3 lm/ft^2 (or 33 lux). An analysis of the visual tasks to be undertaken by a nurse in an emergency indicated that these could be accomplished without serious difficulty in an illumination as low as 3 lm/ft^2 (33 lux). This level of general lighting in the ward will, in combination with suitable values of reflectance of the wall, ceiling and floor surfaces, provide a general luminance level far below the limit for perceptible glare discomfort for people in normal health, and which was therefore believed to be satisfactory for sick people. Such a low level of illumination must, however, be supplemented by additional lighting in the form of bedhead lighting for the patients for their own activities, and for nurses and clinicians for use when necessary. It would be expected that any nursing activity in the general lighting in the ward would be confined to the simplest visual tasks except in an emergency. Otherwise the bedhead lighting would be brought into operation.

When the final details of the system were worked out, it was proposed that the illumination falling on the head end of the bed from all sources not under the patient's control would be kept down to a level of 30–50 lux, whereas the illumination in the central area would be higher than this, of the order of 100–200 lux (IES 1968B). The illumination to be provided by the bedhead lamp under the patient's control (also available for the nurse when attending to the patient), should be not less than 200 lux and could be considerably higher than this.

The distribution of luminances in an interior cannot, of course, be determined without relation to the scheme of decoration of the floor, walls and ceiling. The reflectances of these surfaces cannot be determined from considerations of artificial lighting alone, because modern daylighting technology depends upon the internally reflected component of daylight to provide a major portion of the daylight illumination, especially in areas remote from windows. With daylighting considerations in mind, the Ministry of Health has recommended a standard series of reflectance values for internal room surfaces. These are:

Walls—50% reflectance
Floors—30% reflectance
Ceiling—80% reflectance

Control of Glare

These requirements of illumination levels and distribution, and of room surface reflectances, set the limits within which the detailed design of the artificial light distribution can be specified. In addition, the permissible glare limit in hospital wards must be kept very low. The current IES Code (1968A) recommends a limiting Glare Index value of 13. This strictly limits the maximum luminance of the bright portions of lighting units which might be seen by patients in bed. A maximum luminance of 300 ft–L (approx. 1000 cd/m^2) will under most circumstances be within the IES Code requirements with respect to relatively small areas such as would occur with filament lamp units. With fluorescent units of much larger size, the maximum permissible luminance is in the region of 150 ft–L (or 500 cd/m^2). The values of luminance in relation to projected area recommended in the IES Technical Committee's Report on Hospital Lighting (1968B) are shown plotted in Fig. 10.2.

A typical general ward lighting unit is shown in Fig. 10.3, which uses twin 80 W fluorescent lamps. Such a single fitting is adequate for the general lighting in a four-bed ward of dimensions about 21 ft × 18 ft (6·4 m × 5·5 m).

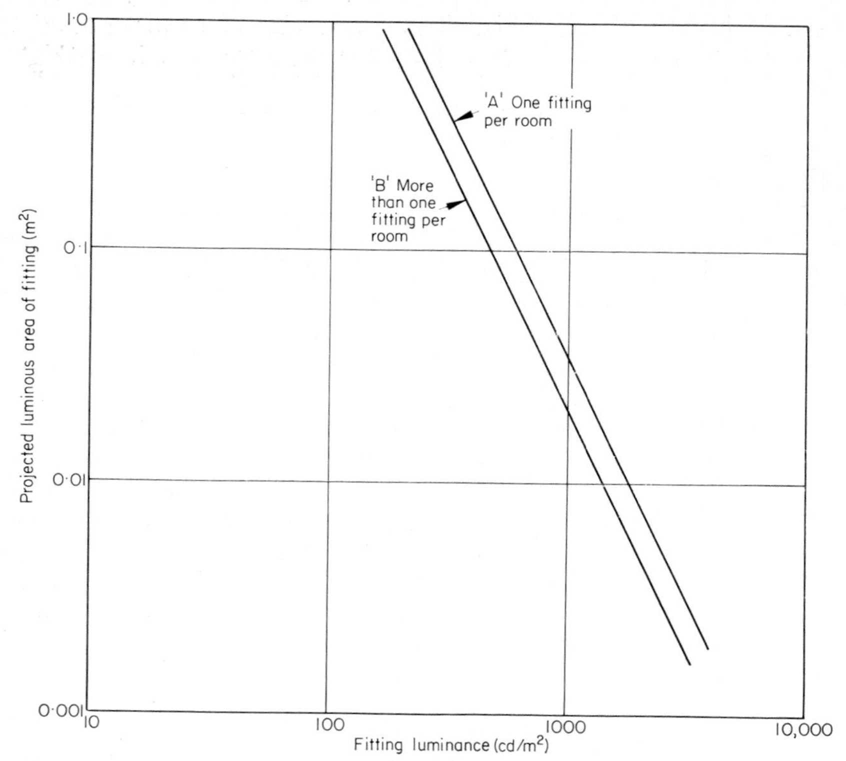

Fig. 10.2 *Relation between fitting luminance and projected area for Glare Index 13 from suspended fittings in hospital wards*

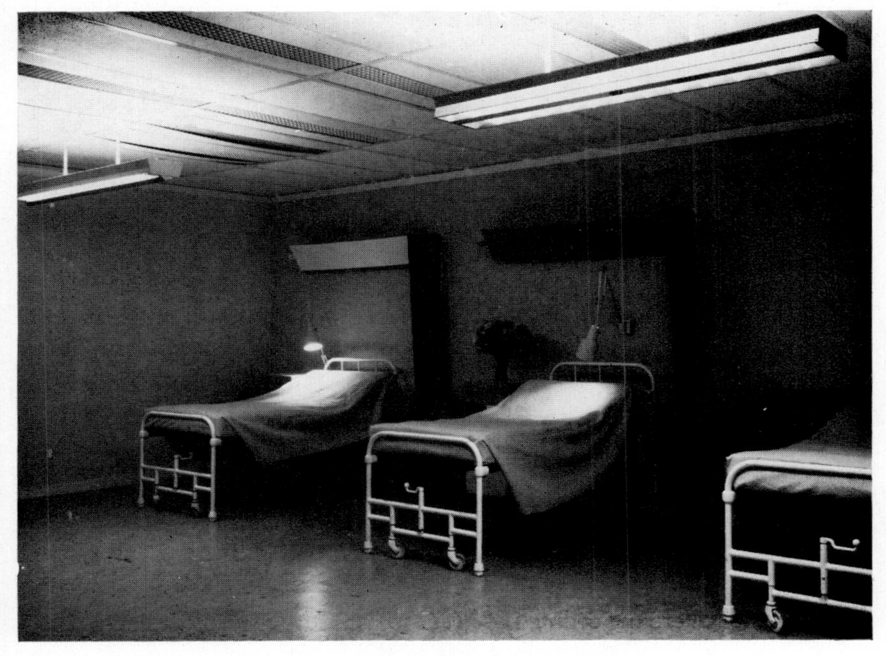

Fig. 10.3 *Evening lighting in hospital ward*

Distribution of Light

The distribution of light from the general lighting units was determined primarily from considerations of the lighting character required in the ward. The lighting can be made deliberately bright and stimulating, or it can be deliberately soporific; it can be institutional, or it can be domestic in character. There are arguments in favour of each, but the Nuffield architects responsible for the original decisions, and subsequently the Ministry of Health, endorsed the view that the lighting should be as far as possible domestic in character, free from unnecessary brilliance and stimulation, but not soporific.

Indirect lighting always produces a depressing and soporific character in an interior. In addition, indirect lighting from the ceiling would not be indicated in a hospital ward where the ceiling is such a significant part of the field of view of recumbent patients. Direct lighting, on the other hand, leaves ceilings in gloomy darkness. A combination of some direct and some indirect lighting, in suitable proportions, can go far to create the domestic character demanded. Most people associate such a character with the light distribution from a domestic standard lamp.

The final distribution which was worked out provides the greater amount of the illumination in the central part of the room by direct lighting from the unit, whereas the greater part of the lighting falling on the bedhead arrives indirectly after reflection from the ceiling and walls of the ward. A limit was set by the Ministry of Health for the upward light of between 20% and 30% of the total light output of the unit. This assumes that the unit will be suspended not less than 1 ft below the ceiling. (Where low ceilings do not permit the use of the indicated suspended lighting units, ceiling-mounted or recessed fittings cannot emit as much as 20% of their light output on to the ceiling, and so the visible luminance of the lighting unit must be further restricted.) Thus about three-quarters of the light will fall directly on areas of the floor and elsewhere in the centre of the ward, but with the relatively high reflectance (30%) of the floor, some of this light will add to the total inter-reflected light in the ward and so soften the total effect.

As explained above, indirect lighting is not desirable. Circumstances sometimes make it necessary to use this type of lighting, and where this must be done, the ceiling luminance must never exceed 50 ft–L (170 cd/m^2) and should preferably be much lower. The distribution of light should be such that, combined with the specified wall reflectances, the luminance of the wall at any point does not exceed that of the ceiling, and the variation of luminance over the wall should not exceed 10:1.

The recent tendency to design hospitals with ceiling heights in the wards of 9 ft (2·7 m) or less makes the use of suspended lighting units more difficult than in wards where the ceiling is 10 ft (3 m) or more. The most likely use of recessed and indirect lighting in wards in the future is in combination one with the other. The need to recess units will carry an obligation for some light to be directed upwards to the ceiling in order to minimise harsh contrasts. This light may therefore have to be provided by separate indirect lighting units.

The final form of the light distribution was determined by subjective appraisals in a fully-detailed scale model of the ward. Observers, including the architects, were asked to express their opinions of the appropriateness of arrangements of the lighting to satisfy the desired criterion of a domestic 'homely' character.

The bedhead lamp for the patient is to enable him to read or to undertake any other individual visual activity, and so the primary requirement is adequate illumination on his work whether he is lying or sitting. (The Ministry of Health recommendation is for not less than 15 lm/ft^2 (approx. 200 lux.) The bedhead fitting should therefore have some adjustment in distance, height and angular movement from the mounting point of the unit, together with a certain amount of adjustment of the shade to tilt forward and back as well as sideways. Any such adjustment must be strictly limited to the amount which is sufficient for the patient's own purposes, but not sufficient to permit under any circumstances the bare lamp or very bright internal surface of the shade to be visible to any other patient in the ward whether they are in the bed opposite or the bed alongside.

A lighting fitting which meets these requirements is illustrated in Fig. 10.4. This fitting has double spring-loaded arms with a parallel link motion which

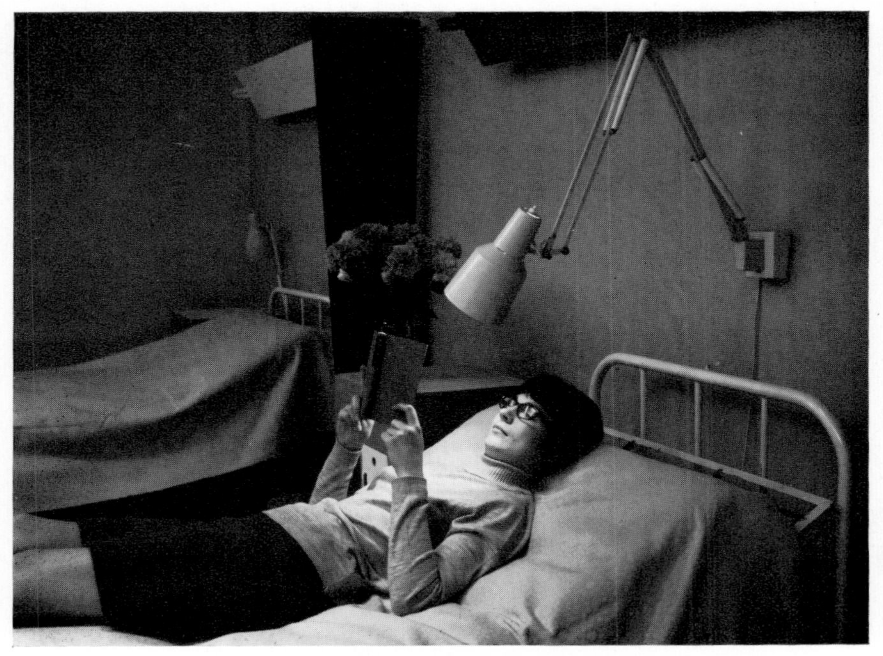

Fig. 10.4 *Patient's adjustable reading lamp*

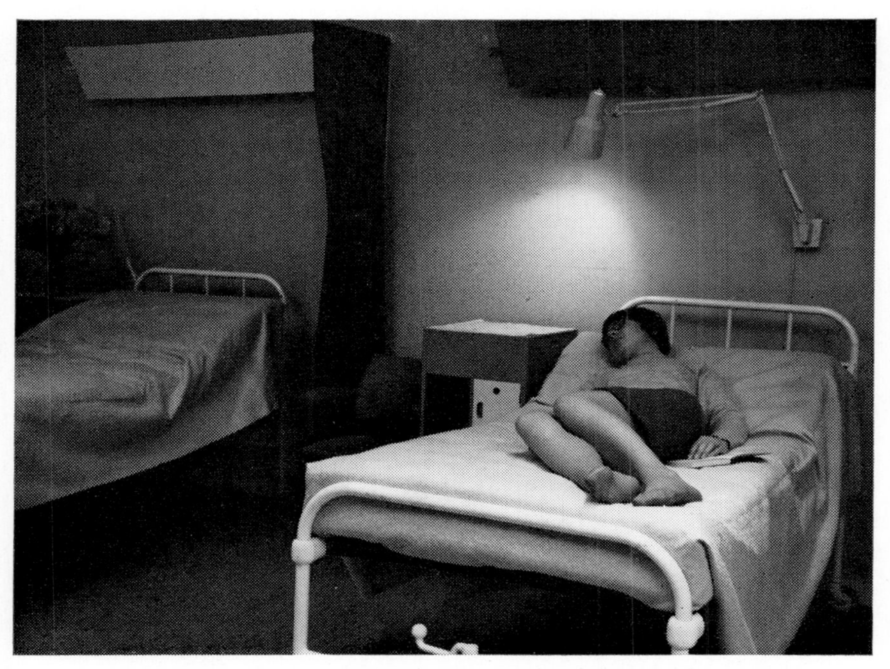

Fig. 10.5 *Patient's reading lamp, showing limitation of movement*

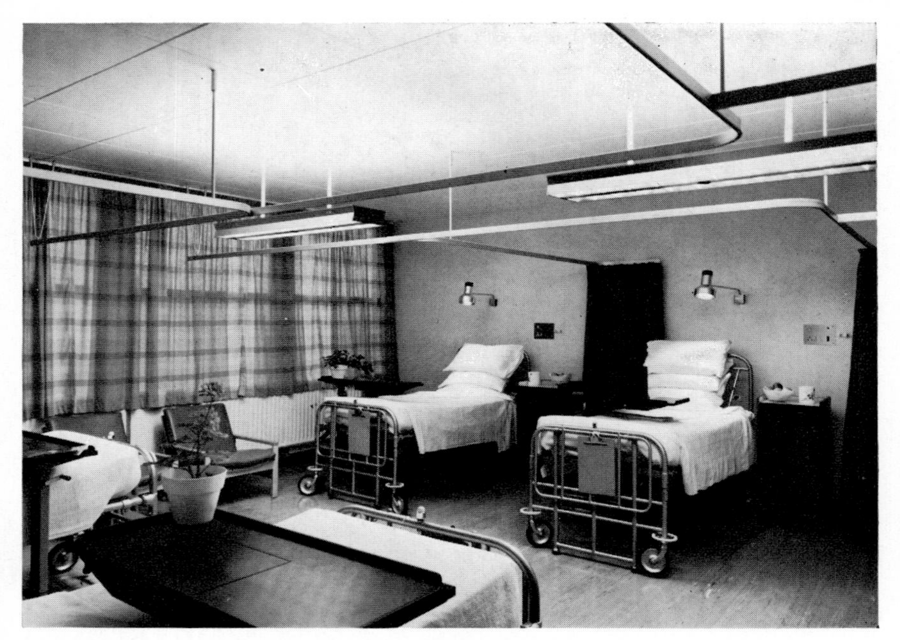

Fig. 10.6 *4-bed patient room at St Mary's Hospital, Portsmouth lighted to Ministry of Health recommendations (Courtesy of British Lighting Industries)*

keeps the axis of the reflector at a constant angle of tilt as it is moved in the vertical or horizontal plane. Special stops on the reflector mounting allow a certain angle of tilt (17°) which, in conjunction with the deep shade, prevents the bare lamp from being seen by other patients however the lamp is misused by the patient, or pushed out of the way by nursing staff when attending to a patient (see Fig. 10.5). The lower 2 in (51 mm) of the inside of the shade is painted to a low reflectance in order to reduce the luminance of this part of the fitting as seen by other patients. Fig. 10.6 shows a complete ward lit to the Ministry of Health recommendations, and Fig. 10.7 gives the luminance values of significant areas of a 4-bed ward illuminated by a single fitting.

Night Lighting in Wards

At night after 'lights out' patients must be able to go to sleep without any disturbance from the general lighting, but at the same time the night staff must be able to move about the ward easily and to see whether a patient is restless or needing attention. Lighting alone cannot achieve all this unless the nurse goes about her business in such a way that her eyes are always more or less dark-adapted. In such a dark-adapted condition, a level of illumination of the order of bright moonlight (0·01 lm/ft^2, or about 0·1 lux) is adequate. Young nursing staff will have no difficulty in adapting quite quickly (in a matter of seconds) to this level after coming into a ward from a more brightly

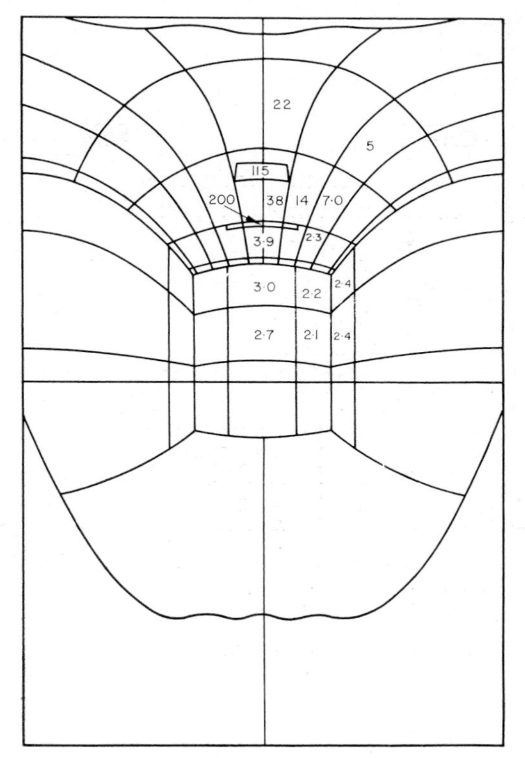

Fig. 10.7 *Luminance distribution in 4-bed ward (ft–L). Two 80 W Colour 34 lamps* [*Illumination on eye* (*without fitting*) *8 lm/ft²*]

lit area, but elderly people may have a little trouble. This level has, however, been recommended as the best compromise by the Ministry of Health as the level of illumination at the bedhead for night-time lighting.

Because of the disturbing effect of bright patches of light on the ceiling in an otherwise darkened ward, it has been recommended that the night time lighting should be provided by screened fittings which throw a pool of light on to the floor between the beds, rather than by indirect lighting from the ceiling, as has sometimes been advocated. The floor then reflects this illumination and redistributes it evenly to the walls and the ceiling, so that only very low luminances are visible to the patient should he be awake.

The practice sometimes adopted of mounting louvered fittings in the walls below bed level to provide night lighting is not considered satisfactory. Lighting the floor under the bed puts the patient in shadow and the nurse sees him or her as an unilluminated object against a bright floor, a situation which is distinctly unfavourable under dark-adapted conditions for making the most of the available light. (A similar situation arises from the use of fluorescent linoleum for night-time lighting, although observations in a mock-up single-bed ward gave the impression that it might be useful in small rooms.) Furthermore, the shadows thrown by the legs of the beds and furniture make a confusing pattern on the floor for the staff trying to find their way about. It is also very difficult to avoid a bright reflection of the lighting unit in a polished floor, as it will be reflected at near-glancing incidence. Consideration should also be given, particularly in children's wards, to the fact that floor-mounted fittings cause 'giant' shadows to move about the ceiling when staff walk around; these shadows can be terrifying to sick children.

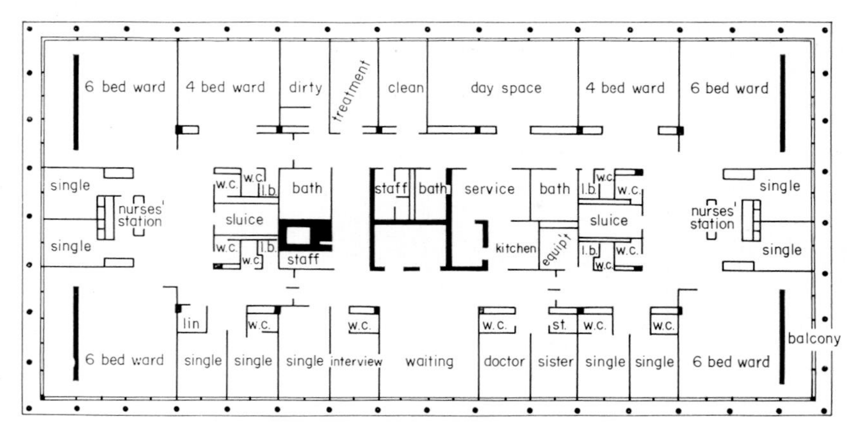

Fig. 10.8 *Plan of wards on a typical floor of a multistorey compact plan hospital block (Copyright: Powell & Moya, Architects)*

Lighting of 'Compact Plan' Hospitals

In a compact plan hospital (Fig. 10.8) a lighting problem of special complexity arises because the central service rooms and corridors have no direct daylighting and depend entirely upon artificial lighting, since these rooms are very close to the wards, staff move from the internal artificially lit rooms through the corridors to the wards in a few seconds and their eyes have to adapt to the different conditions. This type of hospital plan has arisen from a need to make the most efficient use of a hospital site, particularly when the hospital is centrally placed in a city and using very expensive land.

In a building of this type, the three adjacent areas, wards, corridors, and internal service rooms, will have inherently different quantity and quality of lighting. The frequent movement of staff between one area and another requires that the lighting in the three areas should be so graded that when people move from one room to another, they suffer no visual disability or feeling of dissatisfaction due to unfavourable adaptation. During the daytime the wards will be illuminated by daylight and the illumination may vary from around 300 lm/ft^2 (roughly 3000 lux) near the window on a bright day to about one-fifth or one-tenth of this value at the back of the ward. On a dull day, the illumination at the back of the ward may be no more than 5 lm/ft^2 (54 lux). Staff will be moving directly from the ward through the corridor to the internal artificially lit windowless service rooms. The problem under daylight conditions is therefore to know what level of illumination should be provided in the internal rooms in order to avoid any feeling of deprivation or of insufficient light when people come in directly from the ward.

During the evening the problem is somewhat different. The normal ward lighting will be in use and there will then be a much less severe contrast between the illumination in the three different spaces. After 'lights out', however, the ward will be only dimly lit, and the problem is then to ensure that staff who come out of the well-illuminated internal rooms into the darkened ward can see immediately whether all is well with the patient.

The speed of adaptation of the eye from one level of luminance to another has been studied in terms of factors such as threshold contrast visibility, but in the hospital situation it is the subjective assessment of the situation by the people actually working in the hospital which is the determining factor. Basic data on the speed of visual adaptation can predict approximately what will be the most favourable lighting conditions, but the refinements can only be made by careful subjective assessments. In order to obtain such assessments, a complete mock-up of a ward unit was built as part of an investigation by Ne'emann, Isaacs and Collins (1966) at the Building Research Station. This ward unit was intended to be a representative section of a typical 'race-track' plan hospital, and consisted of three parts, a four-bed ward, a short

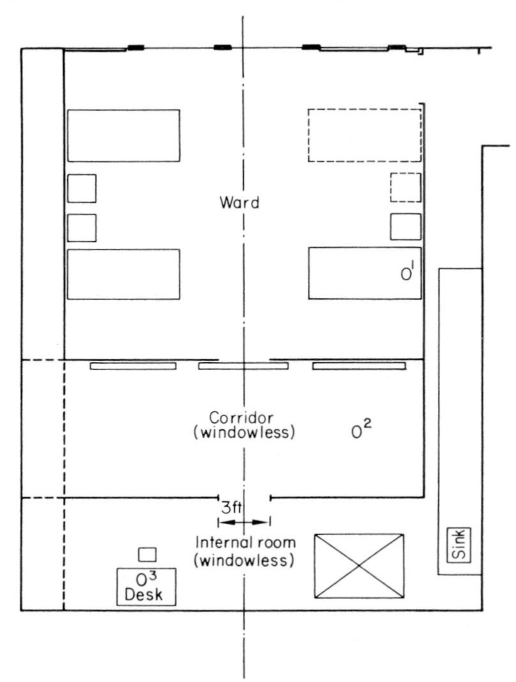

Fig. 10.9 *Plan of BRS mock-up hospital ward*

section of corridor, and a small internal windowless room. The layout is shown in Fig. 10.9.

The glazing in the outer wall was arranged to provide a daylight factor of 1 % on the bed farthest from the window, this being the minimum level of daylight factor recommended in the British Standard Code of Practice (1964) and endorsed by the Ministry of Health. The ward was furnished with beds and bedside lockers and the walls were painted light and medium grey (reflectances 70 % and 55 %) while the floor was covered with grey linoleum of reflectance 30 %. These values are within the recommended limits of the Ministry of Health. The window arrangement is shown in Fig. 10.10.

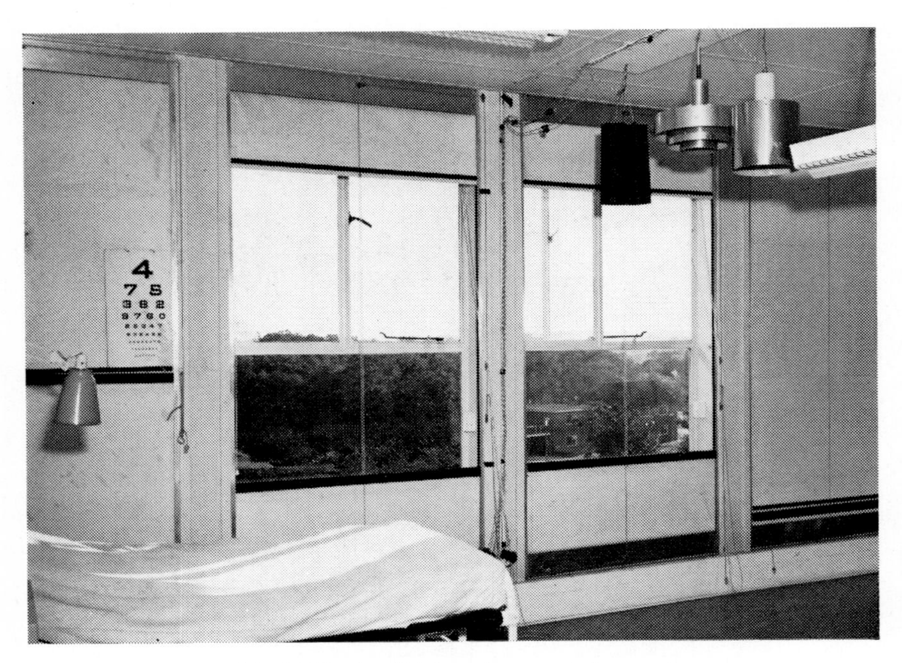

Fig. 10.10 *Daylighting of compact plan ward*

The corridor forms an important visual link between the ward and the internal room. It serves as an 'adaptation passage' in both directions. The corridor lighting should, of course, avoid causing discomfort to patients being wheeled down the corridor; this is probably the overriding factor in the design. If this requirement is satisfied, the next priority is to provide a level of luminance which represents the midway stage between the adaptation conditions in the ward and in the internal rooms. This midway stage is far more important at night after 'lights out' than during the daytime when the situation is much less critical.

The internal windowless room in the experimental mock-up was provided with one light grey wall of reflectance 70 %, while the rest of the internal surfaces were white, except for the floor covering of reflectance 30%. The internal room was illuminated by fluorescent lamps screened with low luminance louvers.

The illumination in the corridor and internal room was adjustable on dimmers with separate control for each space. The ward lighting was set in turn to fixed values consistent with the Ministry of Health recommendations, i.e. 1% Daylight Factor (minimum) for daytime and 0·01 lm/ft^2 (0·1 lux) for night-time.

The basis of the experiment was to ask a team of skilled observers to select the illumination in both the corridor and the internal artificially lit space at such levels that a satisfactory impression of each place was obtained under certain specified critical conditions.

The critical condition for the internal room lighting occurs when the ward is lit by daylight. People coming directly into the internal room from the daylit ward, especially on a bright day, with only the short space of time taken in crossing the corridor to adapt to the new conditions, will be particularly sensitive to inadequate lighting in the internal room. The critical condition for the corridor will be when coming out of the internal room and, looking through the door of the ward, seeing the bright sky through the windows in contrast to the corridor walls.

The illumination level in the internal room was determined by allowing the observers to adapt to the daylight conditions in the ward and then to ask them one by one to go through the corridor into the internal room opposite and to raise the illumination level, initially set by the experimenter at a low level, to give three criteria in turn, as follows:

1. Lighting level just too low for satisfactory appearance.
2. Room adequately lit.
3. Room just too bright.

The results are shown on the histogram in Fig. 10.11. It was found that the level of illumination selected in the internal room did not vary greatly with the prevailing daylight conditions, although on the brightest days appreciably higher levels for the internal illumination were chosen than on darker days. Nevertheless it was felt justified to conclude that a recommendation could be made for the level of internal room lighting which would be satisfactory for all daylight conditions.

The value of asking observers for more than one criterion of subjective assessment is apparent from the histograms. The values of preferred level are spread over a wide range, but when the responses for 'just too low' and 'just too high' are considered, the optimum design values of the illumination level become much more obvious. Thus from Fig. 10.11 it appears that the level of illumination in the internal room should be between 35 and 45 lm/ft^2 (about 380–500 lux) on bright days, and between 25 and 35 lm/ft^2 (about 270–380 lux) on dull days. The Ministry of Health has recommended, on the basis of this evidence, that the level of illumination for internal room lighting

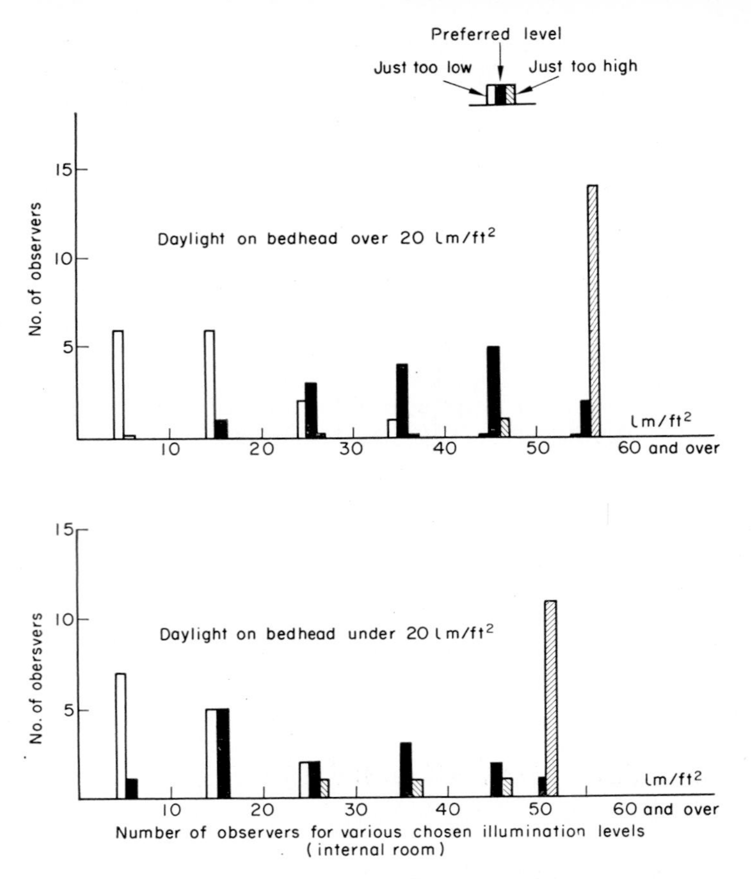

Fig. 10.11 *BRS compact plan ward. Illumination of the internal room*

should be between 30 and 40 lm/ft^2 (about 330–430 lux). Since the preferences are not precise, such levels will be satisfactory on dull and bright days equally.

A similar procedure was used to determine the illumination in the corridor. The Ministry's recommendation here, on the basis of the experimental work, was that a value of between 20 and 25 lm/ft^2 (220–270 lux) should be provided on the centre of the floor of the corridor during the daytime.

The really critical conditions occur at night after 'lights out'. During the daytime and during the evening the wrong lighting in the corridors or the internal room may give dissatisfaction and a feeling of deprivation, but provided levels are in accord with current practice, no actual visual difficulty should result. During the night, however, the staff, when dark-adapted to the moonlight level of the ward illumination (0·01 lm/ft^2 or 0·1 lux) will be able to see quite adequately if patients are needing attention, but if they have spent some time in a service area where the illumination is related to the visual difficulty of the tasks to be carried out in these areas, it will take a certain amount of time to adapt to the conditions in the ward. It is essential that staff should adapt quickly enough to be able to resume observation of patients without a dangerous delay, quite apart from being able to go into a ward immediately and see their way about in safety. Although extensive data exist on dark adaptation, particularly to threshold contrast conditions, little work has been done on the length of time which is taken before people can perform a given visual task in the dark, and no work directly relevant to the hospital problem. For this reason further experiments were carried out in the mock-up ward.

The illumination in the ward was set at the recommended level of 0·01 lm/ft^2 (about 0·1 lux) at the bedhead. The illumination on the floor of the corridor was set at 1·0 lm/ft^2 (11 lux) a figure chosen on the basis of observations made from within the ward under night-time conditions, while the illumination on the working plane in the internal room was set first at 35 lm/ft^2 (380 lux), the mean value recommended by the Ministry of Health for daytime conditions, and later, at a value of about half this (in fact, 20 lm/ft^2 or 22 lux), for repeat observations.

In the course of the experiment, two different aspects of dark adaptation were studied: (a) acuity adaptation, using standard Snellen charts; and (b) contrast discrimination adaptation, using Landolt ring charts. The Snellen charts were of black letters on a white background, or of white letters on a black or a grey background. The Landolt rings were all white, on a light or dark grey background.

Observers spent between 10 and 20 minutes in the internal room in order to adapt fully to the prevailing level of illumination. This was thought to be well in excess of the maximum time which a nurse, who has to return frequently to the ward, would spend in the internal room. At the end of this adaptation period, the observer walked across the corridor, stood just inside the ward, and looked across to one of the test charts placed on the wall close to the bedhead at a distance of 6 m from him and at a height of 1·68 m (5·5 ft). The illumination on the charts when the bedhead illumination was 0·01 lm/ft^2 (0·1 lux) was actually 0·014 lm/ft^2 (0·15 lux), the difference being due to the fact that the light distribution from the suspended fittings (to Ministry of Health recommendations) gave slight preferential lighting on the walls.

The time taken for the observer, after entering the ward, to see and read each line on the chart was recorded; or, in the case of the Landolt ring chart, the time was noted before the observer called out correctly the orientation of the first ring on the chart and also the time taken to call out correctly the orientation of all six rings on the chart.

Fig. 10.12 and 10.13 show the course of dark adaptation. Even when the internal room is lit to the higher level (35 lm/ft^2 or 380 lux), the 6/36 line of the high-contrast Snellen chart can be seen five seconds after entering the ward, and acuity as high as 6/18 can be obtained in less than one minute.

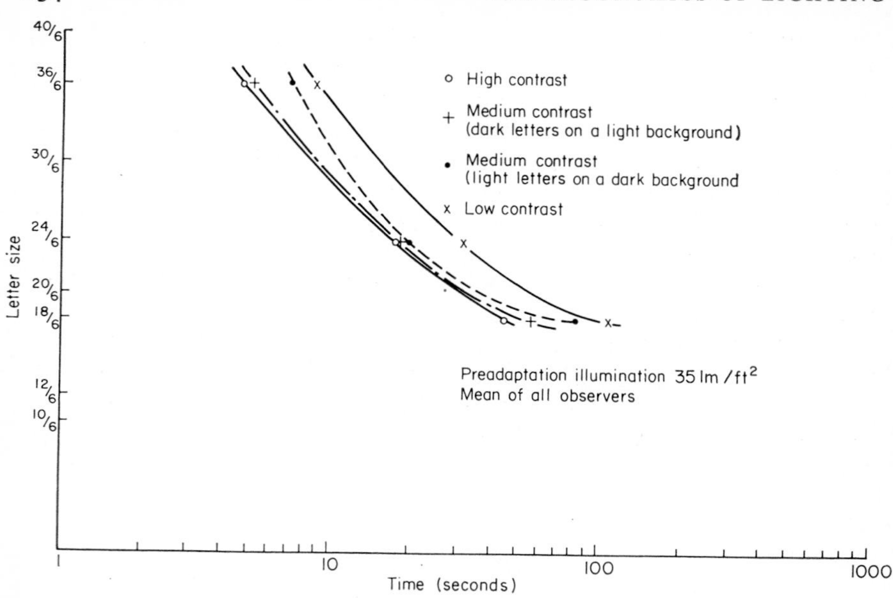

Fig. 10.12 *Relationship between time taken to see lines of letters on a Snellen chart and the size of the letters*

Reduction of the contrast of the chart increases the time. Lowering the illumination in the internal room to 20 lm/ft^2 (220 lux) roughly halves the time taken to adapt sufficiently to be able to see the high-contrast Snellen chart. One of the significant findings, confirming what is already known about the effect of age on vision, was that people in the older (45–50) age groups can

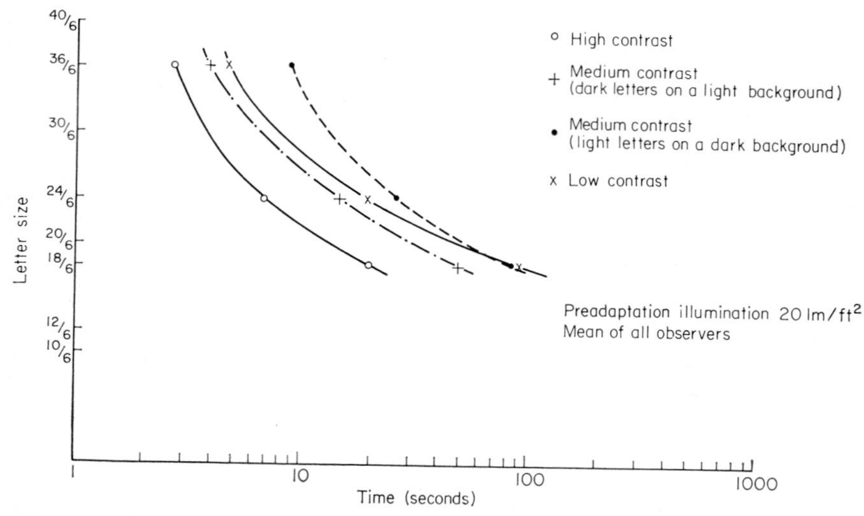

Fig. 10.13 *Relationship between time taken to see lines of letters on a Snellen chart and the size of the letters*

take as much as three times as long to adapt as those in the younger (20–25) age groups. The change of the preadapting illumination does not appear to favour one age group in relation to another.

These experimental data are only indicative, since a very large number of observations from a very large number of people would be necessary in order to make the results definitive. Nevertheless the information is sufficiently in line with what is known from contrast threshold studies to give confidence that application of the results will not lead to trouble in practice.

It was interesting that, on the average, light letters on a dark background took slightly longer to see than dark letters on a light ground. When the preadapting illumination was reduced to 20 lm/ft^2 (22 lux), the difference between the two types of chart was increased. This finding is somewhat at variance with the work of Hopkinson (1949) and other workers, who found that visual acuity under low levels of illumination was slightly better with white letters on a dark background than vice versa. None of these investiga-

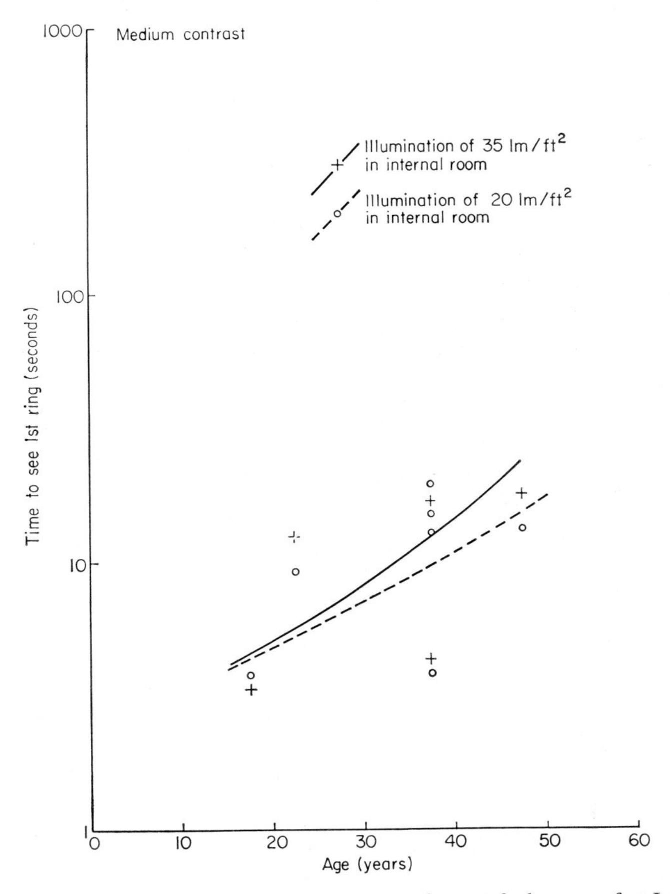

Fig. 10.14 *Relationship between adaptation time and age of observers for Landolt ring charts*

tors, however, made studies down to a level as low as 0·01 lm/ft² (0·1 lux), and moreover they were concerned with final level of visual acuity when fully adapted to the low illumination, rather than with the speed of adaptation as in the present case. The findings may need to be taken into account in the design of signs to be read at low levels of illumination by people of a wide range of ages who have previously been adapted to much higher levels.

The results on the contrast judgments with the Landolt rings followed more or less the same trend. People in the older age group took from three to five times as long to adapt as those in the younger age group, the older people appearing at a greater disadvantage with the higher levels of preadaptation. The results are shown in Fig. 10.14 and 10.15.

The results of both studies showed that if an illumination of 0·01 lm/ft² (about 0·1 lux) was provided at the bedhead, this would ensure sufficient lighting in the ward for staff to be able to adapt to a standard of vision (6/36)

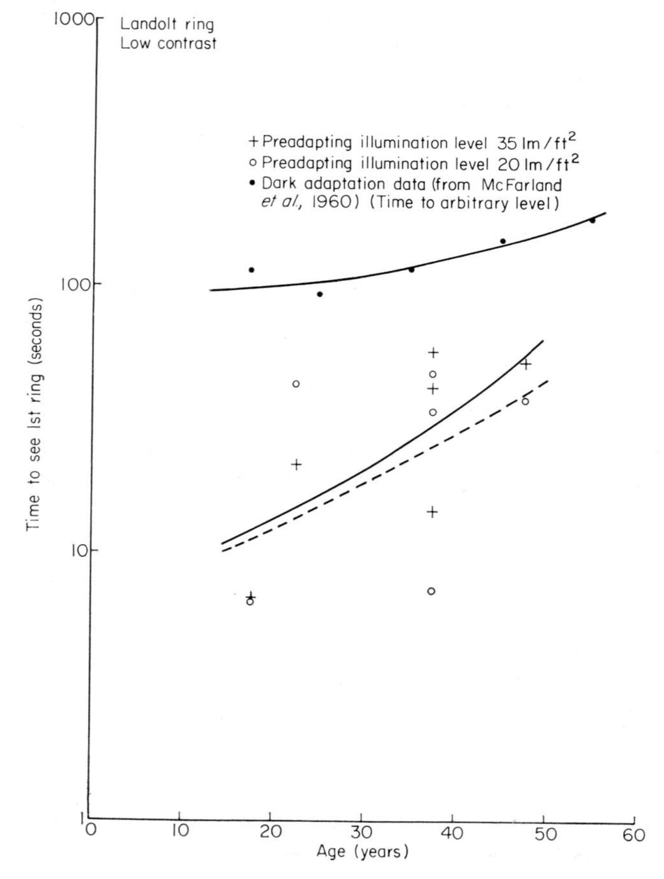

Fig. 10.15 *Relationship between adaptation time and age of observers for the different preadapting illumination levels using a Landolt ring chart*

sufficient for general supervision almost immediately on entering the ward from across the corridor, having left a service room illuminated to 35 lm/ft^2 (380 lux). If the illumination in the service room is lowered to 20 lm/ft^2 (220 lux) during the night, it will have the advantage of reducing the adapting time by about 30%. On consideration, this was thought to be a worthwhile improvement under the conditions in which nursing staff have to carry out their work, especially as the reduction in ease of seeing in the internal rooms will be quite small.

The recommendations for the lighting of wards, corridors and internal rooms in compact plan hospitals, which have been put forward as a result of these experimental studies, are as follows:

1. *Illumination of internal rooms*: During the daytime the illumination on the working plane in internal service rooms should be between 30 and 40 lm/ft^2 (330–430 lux). At night, the illumination in internal rooms so placed that nursing staff may move directly across a corridor into a dimly lit ward should be of the order of 20 lm/ft^2 (220 lux). The level of illumination in the internal rooms can, with advantage, be lowered from the daylight value to 20 lm/ft^2 (220 lux) as soon as external daylight fades, and the artificial lighting is switched on in the ward, with the proviso that local lighting may have to be provided for any particularly difficult task (e.g. reading fine graduations on syringes).

2. *Illumination in the corridor*: The illumination on the floor of the corridor between a ward and an internal room should be adjusted to provide a gradient between the illumination levels in the two spaces on either side. During the daytime, an illumination of between 20 and 25 lm/ft^2 (220–270 lux) on the floor at the centre of the corridor is recommended. During the evening, when the ward lighting is switched on, this value may be reduced to 15 lm/ft^2 (or 150 lux), while at night, after 'lights out' in the ward, the corridor should be illuminated to a level of between 0·5 and 1·0 lm/ft^2 (say 5–10 lux).

3. *Illumination in the ward*: The above recommendations for the illumination in the service rooms and in the corridor depend upon the lighting levels in the ward itself being in accordance with the Ministry of Heath recommendations. These are, that the minimum daylight factor in the ward should not be less than 1%; that during the evening the lighting in the ward should be according to specification, with 3–5 lm/ft^2 (30–50 lux) at the bedhead and 10–15 lm/ft^2 (100–200 lux) in the centre of the ward; while at night the level of illumination at the bedhead should be 0·01 lm/ft^2 (0·1 lux).

References

British Standard Code of Practice (1964). CP3 Chap. 1, Part 1: 'Daylight.'

Collins, J. B. and F. J. Langdon (1960): 'A Survey of Drawing Office Lighting Requirements.' *Trans. Illum. Engng. Soc. (London)*, **25**, 87–114.

Hopkinson, R. G. (1949): 'Studies of Lighting and Vision in Schools.' *Trans. Illum. Engng. Soc. (London)*, **14**, 244–268.

Illuminating Engineering Society, London (1968A). The IES Code: 'Recommendations for Good Interior Lighting.'

Illuminating Engineering Society, London (1968B). Technical Report No. 12: 'Hospital Lighting.'
McFarland, R. A., R. G. Domey, A. B. Warren, and D. C. Ward (1960): 'Dark Adaptation as a Function of Age.' *J. Geront.*, **15**, 149–154.
Ne'emann, E., R. L. Isaacs and J. B. Collins (1966): 'The Lighting of Compact Plan Hospitals.' *Trans. Illum. Engng. Soc. (London)*, **31**, 37–58.

11

Studies of Some Lighting Design Problems: II. Colour; Integration of Daylight and Artificial Light; Windowless Buildings

The Specification of Colour

Colour adds a further dimension to the visual field. The intelligent and rational use of colour as part of a lighting design can assist visual performance and improve visual comfort while also having a special effect on the emotional or psychological state of mind or attitude of the occupants of the building. Visual performance is improved by the introduction of a third aid to discrimination, above those already considered—size of detail and critical luminance contrast. The effect on contrast is obvious. Two areas of the same luminance (brightness) are indistinguishable if they have the same colour, but by introducing a hue difference, they become immediately distinguishable. The effects on visual comfort are no less marked but are often more difficult to quantify. As an example, it is now well accepted that the substitution of green draperies in operating theatres for the traditional white has resulted in a marked improvement in visual comfort. The greater part of this improvement is almost certainly due to the reduction in luminance which results from the use of materials of lower reflectance, but if this were the whole effect, equal satisfaction would be obtained by the use of any other colour or of grey Operating staff and their assistants are apparently convinced that green has an added advantage, and if they are right, clearly this added advantage is associated entirely with colour and not with luminance.

Such little systematic study as there has been, outside the Building Research Station, of colour in buildings has been, however, less concerned with matters of visual efficiency and visual comfort than with the appearance of the building. It is not possible to move very far in the field of colour treatment before becoming involved in questions of current thought and contemporary fashion where there is little that is absolute and little more that is truly rational. The 'scientific' contribution to the problem of colour design in buildings has been largely confined to putting forward precise methods for the specification of colour, through the issuing of warnings to designers on matters of colour 'clash' that can be visually fatiguing or which inhibit good seeing, and to the recommendation of certain simple matters of colour treatment such as the colour of school chalk boards (Hopkinson 1952), matters on which a certain amount of experimental evidence can be offered in support of proposals. The work of Gloag and Medd, which we shall discuss later, however, was founded on visual factors following the adoption of a system of specification which facilitated recommendations for colour design for visual comfort, attention and efficiency of seeing.

The Specification of Surface Colour

There is no scientific reason why the specification of surface colour should not employ the same system as the specification of primary light source colour (discussed in Chapter 7). Nevertheless, it is convenient to use a somewhat different system, particularly because in discussing matters of surface colour, the disciplines of the architect and the interior designer are involved, in which the training in colour language has been entirely different from that of the physicist. Moreover a physical specification does not indicate directly what the colours look like. Experience has shown that surface colour is best prescribed in terms of three main attributes, the hue, the reflectance, and the degree of colourfulness. Of the many colour specification systems which have been put forward, only that of Ostwald and that of Munsell have enjoyed a wide reputation. The system of Hesselgren, which is based on certain concepts of Hering and Johansson, takes into account more variables but is more complex and less easy to manage.

The present discussion will be concerned entirely with the Munsell system. This system, which is now better known than the Ostwald system, has several advantages: (a) any colour reference can be translated directly into physical terms; (b) it can be related to the CIE trichromatic system (Chapter 7); (c) in its present form, it involves no theories of 'colour harmony' which might be thought controversial; and (d) it is not a closed system, as is the Ostwald system, but can be extended to include colours of greater colourfulness and higher reflectance than those available when the system was first laid down. It can also, with relatively little difficulty, permit the specification of fluorescent colours (e.g. 'whiter than white') which have a greater luminance than simple surface diffuse reflectors with 100% reflectance.

In the Munsell system three independent characteristics are used to identify the subjective appearance. These are:

Hue, which is the colour relative to the spectrum or mixture outside the spectrum (e.g. purple) and is the property of the colour which distinguishes it from neutral, white, grey or black. Hue is specified in terms of colour name (red, yellow, green, etc.) combined with a numerical specification.

Value, by which is understood the lightness or darkness of the colour in terms of the amount of light reflected by the surface, and which is therefore correlated with *reflectance*.

Chroma, which is the 'strength' or colourfulness, varying from greyish colours with low chroma to highly saturated colours with high chroma.

Both Value and Chroma are specified numerically.

Neutral grey has no Hue and zero Chroma, but may have any Value between 1 (black) and 10 (white of 100% nominal reflectance). (Fluorescent colours can be included in the system by using numerals higher than 10, but this has not so far been done.)

Fully saturated colours, i.e. those with maximum Chroma, do not all have the same lightness. Thus a red of maximum Chroma cannot have a Value as high as that of a yellow of maximum Chroma. This is an inherent physical property of, for example, reds and yellows.

There have always existed practical difficulties of obtaining surface colouring agents of high Chroma or high Value in certain hues. However, as new pigments are discovered, so blanks in the Munsell system can be filled. This is one advantage of an 'open' system as compared with the 'closed' Ostwald system.

The properties of Hue, Chroma and Value can be used to arrange all known surface colours in a three-dimensional pattern around a natural axis with black at one end of the axis and white at the other. The system can therefore be translated into a practical Colour Atlas which embodies all the available surface colours with known pigments. The published version of the Munsell colour atlas displays over a thousand colour patches (with either matt or gloss finishes) in regular steps of Hue, Value and Chroma. These thousand patches are themselves only a skeleton defining a scale; any given colour in practice might not necessarily prove to correspond exactly to one of the patches in the atlas. Nevertheless it can be defined, either by visual judgment or by colorimetric measurement, to fall into position between adjacent patches in the atlas, and so it can be defined precisely in the three-dimensional colour scale of the Munsell system.

The Munsell scale in the Hue dimension has 100 steps, of which forty are given in standard atlases; in these, capital letters indicate the ten major Hue divisions R (=red), Y–R (=yellow red), Y (=yellow), G–Y, G (=green), B–G, B (=blue), P–B. P (=purple), P–R. The Hue dimension is arranged in a circle and each of the ten major segments in the complete circle is divided into four equally spaced hues. Thus the Hue division letters are preceded by a number, 2.5, 5.0, 7.5, or 10.0, indicating the distance round the colour circle from the preceding main Hue. For example, as Hue changes through pure yellow to pure green, the Munsell steps change in the sequence 5Y, 7·5Y, 10Y, 2·5G–Y, 5G–Y, 7·5G–Y, 10G–Y, 2·5G, 5G. Thus there are eight Hue divisions between pure yellow and pure green. Each of these hue divisions is represented by a separate page of the Munsell colour atlas.

The Munsell Value scale varies in ten steps from very light to very dark.

Munsell Value correlates approximately with reflectance, according to the following formula:

$$R = V(V - 1)$$

where R is the reflectance (%) and V is the Value (Longmore and Petherbridge 1961).

The scale of Chroma is not a closed scale. There are approximately twelve steps from neutral grey (Chroma 0) to maximum possible colourfulness with existing pigments, although some colours, particularly red, extend as far as Chroma = 16. In most Munsell colour atlases, colour patches are presented for each Value step, and for Chroma 1 and Chroma 3 for twenty of the hues, but many additional patches are available in addition to those shown in the atlases.

Gloag (1961) suggests the following subjective attributes to relate to Value and Chroma:

Subjective attribute	*Munsell Value*	*Reflectance*
Very light	9 and 9·25	72%–84%
Light	7 and 8	42%–72%
Medium	5 and 6	20%–30%
Dark	3 and 4	6%–12%
Very dark	1 and 2	1·5%–2%

Subjective attribute	*Munsell Chroma*
Grey	below 0·5
Soft colour	1–3
Medium colour	4–9
Strong colour	10 or more

A full Munsell reference to any colour is given in the order Hue, Value, and Chroma. Thus a pure blue of medium lightness and strength might be denoted by 5·0B5/6 and would be found in the Munsell colour atlas on the Hue page 5·0B, on the horizontal line corresponding to Value 5, and in the vertical column corresponding to Chroma 6.

A Colour Selection for Design—British Standard 2660

One thousand colour samples, as shown in the Munsell colour atlas, is a small selection out of the total number of possible colour variations. Nevertheless the Munsell colour atlas is an unwieldy selection for the practical coordination of colour requirements by users and manufacturers. Because of this, the Royal Institute of British Architects, in cooperation with the Building Research Station, Government Departments and the paint industry, drew up a range of paint colours for building purposes. A small group led by H. L. Gloag of the Building Research Station and D. Medd of the Department of Education and Science, made a selection of colours. With the aid of the Munsell atlas the colours were selected systematically, to satisfy most of the practical requirements for the interior and exterior treatment of buildings, with due regard to the functional aspects of design involving subjective qualities such as coolness, warmth, liveliness and other attributes which are important in this discipline, however difficult they may be to define in physical terms. This range was published as a British Standard (BS 2660: 1955), displaying 100 colours on ten separate cards with the approximate Munsell reference shown beside each colour. Eight of the colour cards each contained colours of one Hue arranged in order of Value down the two edges of the card so that they can be placed close to any given surface which it may be desired to match or to identify. Colours of low Chroma (e.g. soft colours) are arranged along one edge and medium strength colours on the other vertical edge of the card. One card contains the selection of strong colours (colours with high Chroma) while the tenth is a selection of neutral greys.

However, the Munsell Hue, Value and Chroma scales do not describe fully all the colour attributes of most significance in architectural design. This was pointed out by Hesselgren (1954) and led to Hesselgren's Colour Atlas. More recently, work by Evans (1964) and Gloag (1969) has demonstrated that two other attributes are essential to the specification of colour; Gloag's work suggests that these two attributes should be (a) 'power' defining the degree of clarity or freedom from greyness of the colour, and (b) 'weight' defining the subjective equivalence of colours in different hues. The need for these additional attributes can only be described by means of direct inspection of the Munsell colour charts, where it becomes immediately evident that, for example, colours of the same Value and Chroma in the yellow hue range on the one hand, and in the blue hue range on the other, have entirely different attributes. Reference must be made to Gloag's original publications with the associated illustrations.

In co-operation with the Royal Institute of British Architects, Gloag has proposed a new standard range for building purposes which, through the use of these two new subjective attributes, provides a new organisation of colours. It has been found possible to group colours in a scale of five steps from maximum greyness to zero greyness (i.e. 'clear'). Within this grouping the various hues are arranged in the necessary combinations of Value and Chroma to produce sub-groups of colours having equal 'weight'. This arrangement of colours is more directly useful to designers and also offers more direct control over the character of colouring in buildings than was possible by reliance on Value and Chroma alone. The new range has been selected from twelve specific hues rather than from hues representative of ten different broad groups of hue as was the case with BS 2660. This enables the range to include colours which are an exact match in hue at different levels of 'power' and 'weight', the hues being chosen to represent all the major regions and to permit as many harmonious colour combinations as possible.

The new range is not itself intended as a specification standard. It was devised in order to provide an appropriately systematic and comprehensive range of colours for the development of a series of colour specifications for particular materials used in building. Standards drawn from this range would perhaps contain relatively few colours to deal with basic and recurring needs with the assurance that the standard colours chosen for different materials could be directly related to each other.

Colour Design in Buildings

The principles underlying effective colour design in buildings have been given considerable attention at the Building Research Station (Gloag 1961) and by Medd at the Department of Education and Science. The difficulty has been to sort out aspects of colour which are matters of current preference and have no permanence, and matters which are fundamental, which can well include aspects of colour related to the affective states, to 'colour therapy' or to consensus of permanent opinion in matters of colour preference as opposed to transient fashion.

Quantitative Considerations—Reflectance

Purely from the point of view of lighting design, colours of high Munsell Value in an interior serve to obtain a good distribution of internally reflected light, and so the colour treatment may often be determined initially by the photometric requirements of room internal surface treatment for the reflection and diffusion of light (Hopkinson 1954). Once the required reflectances of interior surfaces are defined in this way, colour treatment can then proceed through the medium of the Munsell system. If the reflectance of a wall is defined from photometric requirements, its Value is therefore specified in Munsell terms, and the interior designer can consult the BS colour range for colours of varying Hue and Chroma at the appropriate value. Light colours reduce the contrast between windows, roof lights and lighting fittings and their surroundings, and so contribute to the reduction of discomfort glare where this discomfort is due to contrast (see Chapter 4). Photometric considerations can therefore have a big part to play in the colour treatment of an interior in terms of reflectance.

The ease of seeing a visual task is dependent on its contrast, and the contribution which backgrounds of suitably adjusted contrast in lightness or colour can make to the visibility of many tasks is considerable. In general the background, particularly if it is the bench top or a machine, should be slightly darker than the task and with a somewhat lower chroma than the task itself in order to obtain maximum visual comfort. It is only when items are required to be seen in silhouette, for example, broken threads of dark colour, that light or colourful backgrounds are advisable.

Colour Preferences

Colour preferences depend not only upon the individual, but upon the environment in which he lives. Gloag (1961) believes that 'the normal preference for colour in the interior of buildings in temperate climates is for a balance of colour on the warm side'. This leads to the use of soft, warm colours on walls to balance the coolness of the white areas and the coolness (in factories) of the extensive area of grey metal and other equipment. On the other hand, in 'windowless' types of building, or where the internal temperature may be above normal, a balance of colour on the 'cool' side may be found to be more satisfactory. The interrelation between sensations of warmth arising from ambient temperature and from the colour treatment of buildings is scarcely understood as yet.

The colour treatment of a building can also be used to compensate or to enhance colour characteristics introduced independently of the decoration. For example, in a factory where the work itself involves extensive areas of warm colour, a colour treatment in cool colours may be found more pleasant. In a room where the windows face north and as a result the daylighting is necessarily cool, a warm colour treatment may be more acceptable. Nevertheless, it is unwise to be dogmatic even in these matters. Rival practitioners in the field will argue as cogently for 'enhancement' as for 'compensation'. Some would argue that it is 'dishonest' to change the 'natural' colour balance in a room by some form of compensating colour. This is not the place to enter into these arguments, which rest rather more upon opinion and experience than upon experimental fact.

The introduction of areas of strong colour can give rise to difficulty, because such areas may introduce a distracting or restless effect. Areas of strong colour should be limited to those which can readily be seen, but which do not allow the colour to conflict with other functions of the colour scheme. Similarly, contrasting colours should not be used to distinguish details (other than those to which attention is to be drawn) on a machine, and any differentiation in colour should be one in Value, with preferably no change in Chroma, and certainly no change in Hue.

On the other hand, strong colour can be used deliberately to attract the attention. In a hospital ward, for example, where wall surface colours of high Value are necessary in order to give a high degree of internally reflected daylight, the use of limited areas of strong colour at strategic positions in the interior may be desirable to assist patients quickly to orientate themselves in an unfamiliar environment. This 'chromatropic' effect is similar to the 'phototropic' effect discussed in Chapter 8, and while strong colour may not have quite the same tractive effect on the attention as a bright light, nevertheless the effect can serve a very useful purpose in interiors. Gloag has pointed out that in a windowless interior, orientation is difficult because there are no obvious signs of where the exit might be. He advocates the use of colour to direct attention to the shape of the building, so that the occupants are made aware of their positions in relation to the exits, in order to avoid the risk of panic in an emergency.

Lighting in Factories

Britain continues to suffer from the effects of having been a pioneering industrial nation just as much in its traditional technology of factory lighting as in other aspects of environmental design for industry. The dark and dismal mills of the nineteenth century were multistorey buildings lit by windows on each side, but although these are the buildings most usually imagined when contemplating industrial practice of those days, most industrial processes from the early days of the industrial revolution up to the present day have been carried on in large single-storey sheds with roof lighting. This is a tradition which has continued from the early weaving sheds to buildings on the latest industrial trading estates, and is likely to continue wherever the demand is for large areas of undivided space. In the earliest days, the easiest way of transmitting power from a single factory engine was by belt drive to several floors of a multistorey building, rather than to have an excessively long run of shafting on a single floor. The subsequent development of the small electric motor, which permitted individual drive, eventually removed this advantage, and it was not until land in industrial areas had become very expensive that the costs of building and lighting a multistorey factory could be recovered by a reduction of the ground area required.

The design of roof lighting to provide a uniform level of daylight inside a building raises very few complications. The principles are discussed by Gloag and Keyte (1959) of the Building Research Station, and in more detail by Hopkinson, Petherbridge and Longmore (1966). The greatest amount of daylight for a given area of glass is obtained by the use of horizontal glazing, while the simplest form of roof lighting is by substituting glazing for some of the opaque areas of a pitched 'shed' roof. The drawback to this arrangement is that when the sun shines, unwanted solar heat may be admitted. This may possibly affect the product or process, and it can also make conditions uncomfortable for workers, either by shining directly on them, or by heating up the building to produce uncomfortably high temperatures. Unwanted solar radiation can be limited by the traditional method of painting the glazing in the spring with some form of nonlasting paint or whitewash which will have all been washed off by the rain before the following winter, when the maximum light transmission to daylight is once again required.

The traditional 'north light' roof is a form of saw-tooth sectional construction which admits daylight through the north-facing glazing only, the angle of the glazing being so selected that for the particular latitude no direct sunlight can enter during working hours. North light glazing is perhaps the most widely used of all factory daylighting techniques. It has certain visual disadvantages. Workers facing north will have a large area of sky light in their direct field of view just above the horizontal (see Fig. 11.1), while any

Fig. 11.1 *Facing north in a workshop with north lights*

vertical surface at which they are looking will be in shadow (see Fig. 11.2). If the inner surface of the roof opposite the glazing is painted white, this will help to reflect some light back on to the south-facing surfaces and so reduce contrast, but the only way to reduce contrast satisfactorily is to allow a small quantity of direct daylight in through the south-facing part of the roof. A special asymmetric monitor roof was developed by the Building Research Station (see Fig. 11.3) which is basically a north light system with a small south light sufficiently large to reduce contrast but not to let in a troublesome amount of solar radiation under average conditions.

Most large factories rely upon daylighting for their main illumination. Side windows are only required to provide 'topping up' light at the perimeter of the building. The real value of side windows is to permit workers to see the world outside. Side windows may give no useful contribution to the light on the working plane over the main area of the factory floor since reasonably uniform daylighting can be provided by the roof lighting. The psychological effect of such side windows is, however, considerable. Workers at the centre of a roof-lit area without side windows often complain of a feeling of 'being at the bottom of a well'. Cases can be quoted of factories on trading estates which, while otherwise intelligently designed, are difficult or impossible to let because they have no side view windows.

In a multistorey building, which has no daylight entering through the roof,

Fig. 11.2 *South side of a machine in shadow from north light*

Fig. 11.3 *The BRS asymmetric monitor roof*

the daylight can only be provided through side windows, and therefore it must necessarily be far from uniform in its distribution across the width of the building. Fig. 11.4 shows the extent of the variation in daylight which occurs. In order to provide adequate daylight, there must be a very large area of window glass, and the depth of the building can never be more than three times the height of the ceiling above the working plane, on the assumption that the glazing goes right up to the ceiling. This is a serious limit on design and planning. Such very large areas of glass bring disadvantages in the way of increased internal heat loss in cold weather and unwanted solar heat gain at other times of the year. Glare is also a problem, for even if the sun is not shining directly in through one of the windows, views of large areas of bright sky in the field of vision of workers facing a window will often make conditions visually uncomfortable.

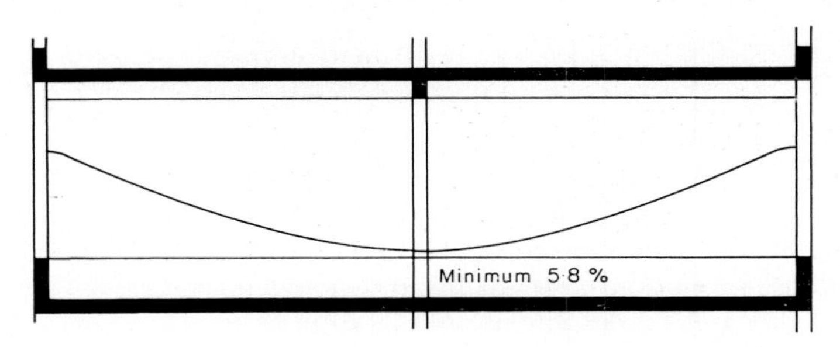

Fig. 11.4 *Daylight factor distribution with side glazing in a multistorey building*
Glass area = 50% of floor area

For these and other reasons it is becoming more common to provide the majority of the working light in multistorey factories by a permanent supplement of artificial lighting. If the working light is provided in this way, windows can then revert to their more useful place of providing a view of the outside and hence can be designed as a part of the visual field rather than as devices to provide working illumination.

Artificial Light as a Permanent Supplement to Daylight

The attitude to the use of artificial lighting during daylight hours has changed radically with the general introduction of fluorescent lighting designed to high levels of illumination. Fluorescent light is cheap to run, and its cost can be balanced against the cost of daylight. New methods of building construction have altered the relative cost of window and wall. The cost of daylight is determined not only as a function of the relative capital cost of windows as compared with wall and the cost of cleaning and maintenance, but also in terms of the restrictions on site planning and use of available floor area which result from reliance upon natural light as the only source of working illumination.

The use of artificial lighting during daylight hours did not arise from any careful economic balance of the relative costs of daylighting and artificial lighting. In fact, once people had at the touch of a switch a level of lighting from an array of fluorescent fittings, of a colour comparable with daylight, they switched the lights on and left them on. This caused concern among management, and various devices to switch off the lights were put on the market, operated by photoelectric cells monitoring the daylight. None of these methods of automatic switching was successful. People found that they liked working with a supplement of artificial lighting, even on bright days.

There are many good reasons why this should be so, but probably the most important is that the provision of a high level of supplementary light brightens up the room and also buffers the glare of the sky seen through the window (see Chapter 4). This buffering effect is most noticeable to people farthest from the windows, who also benefit from a noticeable improvement in the available illumination on the work.

Integrated Lighting

The use of a conventional night-time artificial lighting system continuously during the daytime with conventionally designed daylighting is both wasteful economically and by no means the best solution visually. The technique of permanent supplementary artificial lighting in interiors (PSALI) was developed at the Building Research Station during the 1950s to serve as a logical design technique for the integration of artificial lighting into the lighting of building interiors during daylight hours, as an essential feature of the visual environment compounded of both the daylight and the artificial component (Hopkinson 1957, Hopkinson and Longmore 1959).

The impetus came in the first place from the investigation conducted by Llewelyn-Davies and his colleagues at the Nuffield Division for Architectural Studies, which revealed that the cost per square foot of buildings (in this case research laboratories) varied directly with the room height. This showed clearly the economic advantages to be gained from lower ceilings in buildings, and it could be assumed that equal advantages would be gained in building costs if rooms could be made deeper. Both these changes in room proportions could be achieved provided that it was no longer necessary to rely upon the penetration of natural daylight into the depths of rooms.

During the same period American building design had followed a different path. Due no doubt to the very much more severe and extreme climate, the habit there has grown of excluding daylight (by pulling down the blinds) and relying entirely upon artificial lighting as the working illumination over the whole of the working area.

The concept of PSALI was to develop an integrated lighting technique which would retain the best features of lighting through windows, particularly the extensive view outside, and the best features of well-designed artificial lighting, particularly the controlled environment and the ability to reduce glare discomfort to minimal proportions. Control of sky glare is a much more difficult problem than the control of glare from artificial lighting fittings.

Basic Principles of PSALI

When people are working during daylight hours in a room with full visual access through the windows to the bright sky beyond, their state of adaptation is conditioned, certainly to some extent, by the luminance of this bright sky. This state of adaptation therefore demands levels of interior lighting comparable with those near the window. However, the problem is not merely one of providing these levels of illumination on a horizontal working plane, but rather of providing a luminous environment of the same *apparent brightness* as the areas would be expected to have if they were receiving full daylight from the prevailing sky conditions. The level of supplementary illumination has to be related to the visible sky brightness through the window, and distributed over the whole of the working interior.

PSALI is concerned with the building lighting. Its purpose is to provide a supplement integrated with the available daylighting, as part of the initial design of the building in such a way that daylight and artificial supplement together provide a unity of design in which the daylight character is dominant, but the artificial supplement provides a significant proportion of the working illumination. Although much of the building lighting may come from artificial sources, the design will be successful only if the general appearance partakes of the character of good daylighting, rather than of an artificially lit building with inconsequent windows.

A successful supplementary lighting installation will therefore provide levels of working illumination comparable with those in the parts of the room receiving good daylight, but above all it will provide a luminous environment in the areas which receive less daylight, sufficient to bring up the luminance values to an order comparable with those in the well-lit areas. The purpose of the supplement will particularly be to raise the adaptation level sufficiently to offset the liability to glare from visible skies through windows.

In addition to the requirements of light quantity and distribution, there are also certain stringent requirements for the colour characteristics of the artificial supplement. Ideally the supplementary lighting should have the same spectral distribution characteristics as the daylighting which it is to supplement. Daylight itself, however, varies considerably from the cool light of the blue sky, through the somewhat less blue light of the overcast sky, to the warm light of the direct sun. Consequently one single lamp type cannot produce the spectral characteristics to match all forms of daylight. Ideally the solution would be to select a lamp which matched satisfactorily the type of daylighting considered to be critical for the design. For example, in rooms facing north receiving little or no sunlight, it might be expected that a cooler type of supplementary illumination would be preferred as compared with rooms facing south. This is generally true, but the choice of spectral emission characteristics is limited to the available commercial fluorescent lamps. The most efficient of such lamps, however, have spectral characteristics considerably remote from that of any form of natural daylight. Individual types of lamp may each have characteristics which are considered suitable for particular purposes. This will be discussed later.

Finally, there is the principle involved in relating the lighting during daylight hours with that after nightfall. It can be argued that the character of the lighting after nightfall should be the same as that during daylight hours. Equally there is a strong argument that after nightfall the character of the lighting should change completely, both to give a change in the visual environment and also to provide additional visual aid at the end of the day when people may be tired and not working at full efficiency.

Systems of Permanent Supplementary Lighting

While the basic considerations of supplementary lighting design can be derived from first principles, the techniques now in use for PSALI design have been built up largely around direct experimental studies conducted at the Building Research Station (Hopkinson 1957, Hopkinson and Longmore 1959, Hopkinson, Longmore, Medd and Gloag 1964), from which design data were obtained from subjective judgments made under controlled experimental conditions. These experimental studies demonstrated that choice of the type of supplementary lighting design depended upon the daylight distribution in the interior. A deep room lit from one end only by a single window requires different treatment from a room of more normal proportions or from one lit by windows in more than one wall. Again, the supplementary lighting for top-lit factories requires a different design method from that required for side-lit rooms. The basic principles are the same, in that the levels of supplementary light must be related to the available daylight, and the distribution must illuminate walls, ceilings, etc. in such a way that there is complete integration with the daylight character of the room.

There are two basic systems of permanent supplementary artificial lighting:

1. Localised supplementary lighting.
   (a) Localised lighting directed from the window area.
   (b) Localised lighting coming from the body of the room.
2. Distributed supplementary lighting.

Localised supplementary lighting has as its purpose the provision of additional illumination from artificial sources to reinforce the existing daylighting in some suitable fashion. Localised lighting directed from the window area is intended to add light of the same directional characteristics as the existing daylight, from the artificial supplementary sources. This system of supplementary lighting is in favour on the Continent, particularly in Germany and Scandinavia. The usual method is to mount tubular fluorescent lamps in

Fig. 11.5 *Localised supplementary lighting from window position*

parabolic reflectors above the window head directing light towards the remoter parts of a side-lit room (Fig. 11.5). Advocates of this method of supplementary lighting stress the need for directional lighting in work places. They discount the obvious disadvantages from glare. The method does nothing to counteract sky glare and in fact from certain positions in the room the light fittings themselves constitute an additional source of discomfort. Occasionally the row of fittings above the window head is supplemented by ceiling-mounted fittings a few feet in from the window wall. The system, modified or unmodified, has not found favour in Great Britain.

A preferable form of localised supplementary lighting is to mount the units in or near the areas of inadequate daylight illumination and to provide a quantity and distribution of illumination which supplements the available daylight. The available daylight is therefore supplemented in quantity but not in directional characteristics. By careful placing of the supplementary sources, the inadequate modelling given by light from the windows can be improved. On the other hand, by incorrect placing of the supplementary sources, modelling can be confused and the result may not be satisfactory. Lynes and his colleagues (1966, 1968) have shown how the flow of light in an interior with supplementary artificial lighting at the rear can be made consistent, and Jay (1968) has shown the extent to which daylight can provide the required modelling.

A particular form of localised supplementary lighting is to place a large lay-

Fig. 11.6 *Localised supplementary lighting remote from window*

light of low luminance with a wide light distribution, as shown in Fig. 11.6, over the areas remote from the windows. This method is particularly useful in rooms where economy requires great depth to the room without the increase in ceiling height which would be necessary to secure adequate daylight penetration. The method succeeds if the laylight is distributed over a sufficiently large area to avoid any apparent dark area between the windows on the one side and the supplement on the other. There should be no marked dip in the illumination level over the working plane nor in the luminance of the walls, and the eye should be able to scan the whole room without being conscious of any area being apparently starved of light.

Distributed supplementary lighting is less subject to design difficulties as compared with localised systems. Among the advantages of distributed supplementary lighting is the obvious one that the same lighting system can be used after dark as well as acting as the supplement during the day, provided that certain modifications are introduced. Other advantages include the possibility of providing a smooth gradation of illumination and brightness over the whole working area.

One of the disadvantages of distributed supplementary lighting is the higher initial cost. This tended to operate against the advocacy of this system in the early days of PSALI. Another disadvantage is the difficulty of designing the system to look like part of a daylighting installation. In the event, neither of these disadvantages has been shown to be significant in practice. Current

Fig. 11.7 *Distributed supplementary lighting in model room*

practice in artificial lighting now accepts costs per unit area of the same order as that required for distributed supplementary lighting, and the extra cost of selective switching and other refinements is minimal.

A satisfactory system of distributed supplementary lighting is obtained by building the lighting units into the ceiling as shown in Fig. 11.7 and 11.8. It is advisable to arrange the lighting circuits so that selective switching can be practised to permit some of the units to operate during the day while others are brought into action after dark. Such a lighting system differs little from well-designed artificial lighting and presents few additional complications in design. The chief requirements to ensure good blending with the daylighting are: (a) suitable choice of lamp; (b) suitable screening or louvering of the lighting units to avoid any direct view of the lamps; and (c) arrangement of the units in position and switching to give a satisfactory distribution of illumination and a wall luminance to integrate with the daylighting.

Levels of Supplementary Lighting

The amount of light provided from the supplementary installation should be that which would be given by good natural lighting. At any one instant the level of general luminance in the room should be related to the visible brightness of the sky in such a way that no glare discomfort from the sky results and

Fig. 11.8 *Ceiling and wall lighting in PSALI installation*

the brightnesses in the parts of the room receiving artificial light balance with those parts of the room receiving a full complement of daylighting.

It was established experimentally by the Building Research Station that the desirable level of supplementary light should be linked to the prevailing sky conditions. Ideally this would require a smooth system of light control to vary the level of the supplement with the variation of daylight outside. The control would not be based simply on the physical illumination from the sky but would need to incorporate a certain degree of feedback to take into account the lag in adaptation characteristic of human vision. Eventually such suitable light control systems may be an economic possibility in PSALI installations. The Building Research Station experiments also established, however, that a fixed level of illumination corresponding to the average prevailing sky conditions would normally be satisfactory. This level would normally be of the order of 500 lux (or 50 lm/ft^2) with a sufficiently wide distribution not only to permit this quantity to fall on the horizontal plane, but also to supplement the daylight on vertical walls to approximately the same level.

In the case of supplementary systems which employ recessed fittings, some additional directional lighting from the sides is desirable to soften the strong vertical modelling. During the daytime the horizontal flow of daylight from side windows will combine with the downward flow of light from the ceiling-recessed artificial lighting to give a satisfactory appearance, but at night, when the whole of the working light is coming downwards from the fittings at angles close to the vertical the effect will not be satisfactory. The desired effect is achieved in the example shown by lighting parts of the walls, in this case to accent display areas (Fig. 11.8).

Colour of Supplementary Lighting

Several factors have to be taken into consideration in the choice of lighting to act as a supplement to daylight. First, the colour of the visible areas of the supplementary lighting units must look sufficiently close to that of the available daylight so that the attention is not distracted by the obvious artificiality of the supplementary system. A second consideration is that the colour rendering of the objects in the room must be sufficiently close to that of the natural lighting to ensure that colour judgments made in different parts of the room are not confused.

Choice of supplementary lighting for use by night and day requires that the interior should not look cold and gloomy at night because a bluish source of light has been used to give the best daytime match. At night most people expect a warm light of the traditional incandescent type.

Certain types of fluorescent lamp come near to meeting the somewhat conflicting requirements, provided that the maximum efficiency of conversion of electrical energy into visible light is not demanded. Lamps of good colour rendering properties are less efficient in terms of energy conversion than those which emit most of their energy in a relatively narrow region of the spectrum near to that of the maximum sensitivity of the human eye.

During recent years measurements of the spectral distribution of different phases of natural daylight both including and excluding sunlight have been carried out with the aim of arriving at a series of standardised spectral distribution characteristics to represent these different phases. One standardised distribution has been chosen by the British Standards Institution (see Chapter 7) to specify a source of artificial daylight for colour matching purposes. This distribution is representative of daylight plus sunlight having a correlated colour temperature of 6500 K, which is one of the most commonly occurring phases of natural daylight. This distribution has been adopted internationally (CIE 1968) as 'representative daylight'.

The Building Research Station determined experimentally the most satisfactory colour of fluorescent lamp for use in supplemented installations, as evaluated subjectively in a room which combined side daylighting with the supplementary lighting from the ceiling. The most satisfactory colour to provide a total illumination of daylight plus artificial light of 430 lux (40 lm/ft^2) was one which emitted light having approximately the appearance of a full radiator at 6500 K. This experiment was carried out as a pair-comparison test using a model in which the sources of artificial light (producing the same total illumination on the working plane) could be changed quickly so that the observer could judge, with a short time interval between, which of a successive pair of sources was preferred as giving the best blend with daylight. This experimental finding has subsequently been confirmed in field studies.

A satisfactory appearance during the daytime must not be obtained at the sacrifice of a good appearance at night. However, a light source designed to comply approximately with the British Standard for Artificial Daylight (BS 950:1967) has a spectral distribution close to that of a full radiator and thus contains more red light than the old 'colour matching' type of lamp. Consequently a much more acceptable colour rendering is produced with this type of lamp for night-time use, and the impression of gloominess characteristic of the earlier lamps when used at night is no longer so troublesome. It has been observed in recent lighting installations that provided a satisfactory warm and cheerful scheme of decoration is employed, the appearance of the interior is fully acceptable when high levels of illumination (over 400 lux, or 40 lm/ft^2) are provided at night with the Artificial Daylight standard. There is some evidence that the colour rendering is in fact preferred to that of lamps of lower colour temperature. There is every indication, therefore, that the position will soon be reached, if it has not already, where the lamps which are preferred for supplementing daylight will also be preferred for the night-time lighting, at the appropriate lighting levels.

It is not possible to achieve high quality of colour rendering without some sacrifice of the efficiency of light production. Eventually ways may be found of providing satisfactory colour rendering with higher luminous efficacy, but it is likely always to be true that the amenity of good colour rendering must be balanced against the higher cost of capital installation and of running costs. The choice, however, must be made by those who have to weigh the attractiveness, the pleasantness and, in the case of special interiors like hospitals, the efficiency of the interior as judged by those working in it, against the increased cost. This is in any case in total a very small part of the cost of the building, amounting usually to only 1% or slightly more.

Summing up, the present position is as follows:

1. The best form of artificial lighting which blends most satisfactorily when used for supplementing daylight is the British Standard 'artificial daylight' type of lamp with a correlated colour temperature of the order of 6500 K. Such lamps have good colour rendering properties, and give a match with daylight which is preferred to all other artificial sources. At night-time such lamps give a cool appearance to the interior, which is a disadvantage if the level of illumination provided is less than 40 lm/ft^2 (440 lux), but which causes little or no comment at higher levels. The best forms of modern lamp are, in fact, sometimes preferred over others even for night-time lighting, if the decoration of the interior is appropriately designed. Such lamps have a relatively low luminous efficacy and therefore their use leads to higher initial installation and running costs for the same illumination level (but see Chapter 7, p. 162).

2. In certain special areas, particularly the clinical areas of hospitals, where colour rendering properties may be linked to certain specific colour judgments, other forms of lamp may be advisable. The Ministry of Health recommends a particular type of lamp with a correlated colour temperature of 4000 K, and a high standard of colour rendering accuracy. Such lamps are found to be very satisfactory for general lighting after dark, but during the daytime their slightly warm appearance draws attention to their presence as the supplement to the available daylight.

Experimental studies have shown that lamps of this kind are preferred by clinicians for the judgment of colour for a wide variety of hospital requirements. Such lamps are indicated for supplementing daylight where these special colour rendering requirements are necessary. Such lamps may also be the preferred choice in interiors where the daylighting plays a smaller part over the whole working period than does the artificial lighting, such as in shift working. In such situations matching available daylight may not be the critical design criterion.

3. The common 'daylight' type of fluorescent lamp has a higher luminous efficacy than either of the preceding types. It is possible to select a lamp of high luminous efficacy which gives light of a colour not conflicting too violently with daylight, to be used in situations where initial and running costs are of paramount importance. This is often considered to be the case in industrial areas where colour rendering is of secondary importance. In many factories where there is continuous use of artificial lighting throughout the whole 24 hours, as a result of shift working, and where there may be relatively little daylight which can be directly compared with the artificial supplement, the use of such efficient 'daylight' types of lamp clearly confers many advantages. There are many different manufacturing varieties of this basic type of lamp, with different degrees of improvement of colour rendering accuracy in inverse relation to the luminous efficacy, so the choice has often to be made in relation to the specific requirements of the situation.

Some recent work at the Building Research Station (Cockram, Collins, and Langdon 1970) suggests that the high-efficacy lamp in the range of daylight colour is a reasonable compromise for use in the daytime in offices where colour rendering is not very critical.

4. If the supplementary lighting system is designed in the form of a distributive system, part of which is in use during the daytime and part for additional use at night, the daytime supplement can be provided by the 'artificial daylight' type of lamp to integrate satisfactorily with the daylight, while the night-time addition can consist of 'warmer' lamps (possibly of the de luxe type), so that the combined result is more pleasant and satisfactory at night. These recommendations on the choice of lamp for the artificial supplement must, of course, be made specific with relation to commercial types of lamp which may be existing at the time.

The Window in an Integrated Lighting System

An integrated system of daylight and artificial lighting should be designed from the start by the architect of the building. The windows and the artificial lighting units are two parts of the same system.

In an integrated system, windows need not be designed around their function as a means of admitting some working illumination but primarily to create the architectural character of the space, and to permit an adequate view of the world outside. It is the last of these functions which necessarily has to be subordinated in a design for daylight working illumination alone, and which can be reinstated to its major importance in an integrated design. The negative aspects of window design, the elimination of sky glare, of internal heat loss, and the reduction of unwanted solar heat gain can also be given proper attention. Moreover, one of the major disadvantages of very large windows in city buildings, i.e. the amount of external noise which penetrates through large areas of single glazing, can also receive proper attention in an integrated system.

Experimental studies at the Building Research Station showed that in a supplemented installation, the window area can be substantially reduced below that which would be considered necessary for full daylighting. A more extensive experimental approach by Ne'eman (1969) has shown that there is a limit to the reduction of window area which is acceptable in a supplemented installation, and that this critical minimum size depends upon many factors, one of which is the amount of visual interest in the view outside the window. Ne'eman found that if the window gives out on to a uniform sky, the choice of critical minimum window size is entirely different from that chosen in a built-up area where there is visual interest.

The function of the window in an integrated installation is threefold. First, it supplies working light on a horizontal working plane over some part of the room. Second, it provides a view outside, and this is important. Third, it supplies light to assist in modelling. In a side-lit room, for example, areas remote from the window will receive little illumination on the horizontal working surface, but the daylight on the vertical surfaces assists in modelling and counteracts any dull effect of light coming exclusively from above. This effect is illustrated in Fig. 11.9 and 11.10 where the same room, seen from the same viewpoint, receives light, in Fig. 11.9 entirely from the artificial lighting supplement in the ceiling, and in Fig. 11.10 from both supplement and windows. The modelling effect of the component of daylight on vertical surfaces is most marked, although it cannot be illustrated effectively in monochrome on a photograph because the effect is one of solidity which requires three dimensions for its effective illustration.

These last two photographs were taken in a model which was used to study the subjective aspects of window design for a deep room (Hopkinson *et al.* 1964). The Ministry of Education (as it then was) was concerned to examine the possibilities of freeing the design of deep teaching spaces used for labora-

Fig. 11.9 *Downward modelling from artificial lighting*

Fig. 11.10 *Horizontal modelling added by daylight*

tories, etc., from the restiction that the lighting requirements had to be met by daylight alone during the daytime. Such a freedom would enable a number of deep spaces to be assembled together in a compact block building rather than in the rambling layout which is necessary to allow adequate daylight to enter on at least two sides.

The economies in building in this way are apparent, quite apart from the advantage of having increased wall space for storage and display, and studies were made at the Building Research Station to determine the minimum window area to give satisfactory lighting quality, on the assumption that the necessary working illumination would be provided by the artificial lighting.

The experiments in a model room 24 ft wide × 48 ft deep × 10 ft high (7·3 m × 14·7 m × 3 m) led to the conclusion that an acceptable visual environment, from the point of view of light distribution inside the room and of visual contact with the outside, could be obtained with a window of area as small as 5% of the total floor area. As shown in Fig. 11.11, the modelling was quite adequate with this amount of side-lighting in relation to the overhead lighting. By carefully detailing the design of the window and its immediate surround, and locating one edge close to a flank wall, the glare resulting from the high contrast of the sky brightness with the room surfaces was minimised.

The data obtained from these studies was applied in the design of a model for a new school chemistry laboratory at Southfield, Oxford (see Dept. Education and Science, 1967) and the window design finally achieved and incorporated in the building is illustrated in Fig. 11.11. The main window is adjacent to the main display area and chalkboard wall, and the spill of light on to this provides a luminance link with the outside which is achieved in this way, and in the detailing and colouring of the surroundings to the windows. A secondary window at the back of the room in one of the long walls was inserted in order to provide a measure of cross-lighting (Fig. 11.12), the total window area being 1/15th of the floor area.

The quality of daylight on a vertical plane facing the window is quite adequate on most days to give the required degree of modelling, but its contribution to the illumination on the horizontal working plane is negligible at any distance away from the window. Thus it is immediately apparent on most days that the room cannot be adequately lit by daylight alone and hence the artificial lighting is permanently in use. The appropriate design of artificial lighting installation was determined by further subjective studies in the model of the final design of the laboratory, and it was found that to make relevant judgments on this it was essential for the whole model room to be fully detailed; such items as the white-painted shelf edges and laboratory glassware could all be seen to contribute to the visual character of the interior. It was felt that in order to make the artificial lighting fittings as inconspicuous as possible it was necessary to recess them into the ceiling and use light grey louvers (Munsell Value 8) with their lower edge level with the ceiling plane.

Fig. 11.11 *PSALI in chemistry laboratory at Southfield School, Oxford: main window*

Fig. 11.12 *PSALI in chemistry laboratory at Southfield School, Oxford: secondary window*

In order to establish the optimum level of illumination required from the artificial lighting, a group of observers were asked to adjust the level of illumination inside the model to reach three criteria of satisfaction under a wide range of differing external daylight conditions. It was found that an illumination of 25–30 lm/ft^2 (approx. 300 lux) on the bench tops would be quite satisfactory under most conditions of external daylighting, and in the final building it was decided that the aim should be to make this range of illumination the minimum level to be found in service, so that the initial level of illumination would be much higher.

The design had allowed for the row of ceiling lighting fittings near the windows to be turned off in daytime, but in practice it has been found more practicable to have the whole installation in operation in the daytime as well as during darkness. The final result of this may sound like simply using the night-time artificial lighting during the day, but in fact what has been done is that the design of this artificial lighting has been considered in conjunction with the daylighting design (in particular the design of the windows, their surrounds and positions in the room) specially for use during daytime, and its use at night has been examined quite separately.

One important factor which was apparent from the study of the lighting installation at night was that without daylight to provide the horizontal modelling, the restricted light and luminance distribution from the ceiling fittings alone gave a rather gloomy impression. Fluorescent-lamp floodlights were therefore provided to illuminate the chalkboard and pin-up boards to relieve the monotony of unidirectional (downward) lighting with relatively dark walls, and to reintroduce (by means of the reflected lighting) a small horizontal component into the flow of light (see Fig. 11.9).

The colour of the lamps used ('colour matching' type) was chosen on the basis of previous experiments to be a good match with daylight, and was found to have such good colour rendering properties that any suggestion of 'coldness of light' could be avoided by the colour treatment of the interior. This lamp was found by trials in the final building to be the preferred one for night-time use as well as daytime, in view of its 'clarity of colour rendering' and 'crispness' of apparent source colour.

With skill, the architect can produce particularly pleasing lighting effects from an integrated window-and-supplement system with less difficulty than by the use of side windows with top lighting. On the top storey of a building, or in a single-storey building, it may be desirable to combine side windows with top lighting partly from roof lights and partly from an artificial supplement.

One of the virtues of integrated lighting is that it can be truly integrated with the structure of the building and need not use conventional lighting 'fittings' either suspended or recessed into the ceiling. For the lighting of factories, for example, the artificial supplement can be built into a roof lighting system so that either the one or the other, or both, can be brought into use as required.

Integrated lighting leads inexorably to the question, why have the daylighting at all? Cannot the artificial lighting, built into the roof, perform the lighting job fully adequately, without the inconvenience of window cleaning, the loss of internal heat in winter and the unwanted gain of solar heat in summer, and, perhaps, the annoyance of aircraft noise transmitted by the single sheets of glass in the roof? Why not a fully insulated artificial environment?

Windowless Buildings

Arguments have been put forward in favour of windowless buildings on economic grounds, but they are difficult to sustain over an extensive range of conditions. It is true that a well-insulated building will lose less heat in winter than one with windows, and that maintenance costs may be less because there are no windows to clean, but against this is the cost of the permanent lighting equipment and its running costs. The economic balance will be different in different climates, and will depend upon the cost of power for lighting, for heating, and upon labour costs for maintenance. The proper costing of a windowless building as compared with a windowed building performing the same functions is a specialist exercise.

Such an exercise in any case cannot take into account the psychological factors associated with working in buildings without windows. These may have important economic consequences if they involve difficulties in recruiting and holding labour.

These factors are also of significance so far as the lighting problem itself is concerned. This problem can be stated simply—to what extent should the lighting of a windowless building differ from a conventional artificial lighting system? Apart from the interest from a purely lighting point of view, the answer has economic consequences because it will affect the balance sheet if special lighting at additional expense has to be provided.

Among the environmental arguments for windowless buildings, two are: (a) to control the environment within close limits of temperature and humidity because of the exigencies of the manufacturing process, e.g. in textile or transistor manufacture; and (b) to provide a comfortable working environment for people throughout the year in an unfavourable climate, e.g. Alaska or Siberia; or Arizona or Central Australia. A third reason, recently advanced in North America, is that the absence of windows reduces the risk of damage through rioting or vandalism.

The psychological factors which operate in windowless buildings are not fully understood, and the evidence is full of contradictions. There is no doubt that people do not in general like to work in a building without windows, and as a result it is accepted that 'good natural light' is an amenity which justifies an increase in rental, while absence of windows is an accepted argument for additional remuneration of workpeople. The degree of deprivation is related to the nature of the work and of the environment; it is less in places like underground railway stations where there obviously cannot be any daylight, or in department stores where there is an unimpeded path to the daylight if the worker feels claustrophobic (but she never does when she knows there is nothing in her way); greater in places where there seems no good reason for cutting out the daylight except profit to the management, or where a boring job has to be done which permits the attention to wander.

Claustrophobia, or the expectation of it, is probably the greatest psychological argument against the windowless building, and is the factor which lighting primarily has to set out to alleviate. The secondary factor is the need to provide an internal environment which does not give rise to any short-term or long-term effects on vision.

The second factor can be dealt with quickly. There is no evidence, and there is no reason to expect any, that conventional artificial lighting to an adequate level has any long-term effects on human vision, and it certainly has no short-term effects. Medical staff who have the responsibility for welfare in windowless buildings should certainly be on the alert for any visual difficulties which may arise, but there is no evidence up to the present that there are any associated specifically with the lack of windows. (Problems of flicker, glare, etc. have been discussed, and these arise in any artificially lit building.)

The first factor, the problem of claustrophobia, or simply the dislike of working without a view outside, merits attention. Among the ways in which such effects can be alleviated by lighting or other visual means are:

1. The provision of false or symbolic windows.
2. The creation of the illusion of a window.
3. The use of mural designs which suggest a view.
4. The use of large glass-partitioned spaces within a large windowless building to give the greatest possible viewing distance.
5. The use of special light distribution to give an apparent increase to the heights of ceilings or the distances of walls, or to increase the apparent size of an enclosed space; or the use of special colour treatment in conjunction with light to achieve the same objective.

Symbolic windows are frequently advocated but are not much used except in basement restaurants, where in conjunction with the rest of the decorative treatment a pleasant effect can be created. In workplaces, a useful effect can be obtained by placing symbolic windows above eye-level, where they can attempt to create the effect of a window giving out on to an overcast sky. Below eye-level the effect is lost because the expected view is not there.

The illusion of a window can, however, often be created with advantage. One of the pleasant features of a building with windows is the manner in which sunlight can stream through a window into the room, even though the window itself may be hidden by a return wall or by furniture in the room. This illusion can be created in a windowless enclosure, and it is an illusion which seems to remain even though the artifice is well understood. An extension of this device is to provide a number of alcoves down the length of a windowless space, lit to a very much higher level of illumination with light of the colour which sunlight would be expected to be by the eye adapted to the light of the main room. This colour may not be identical with real sunlight. The choice of colour needs care, but it may not be too difficult; for example, in a main room lit by 'daylight' fluorescent lighting, the use of 'white' lamps, which are of a 'golden' colour by comparison when the eye is adapted to 'daylight' lighting, may provide the desired contrast in colour. These alcoves have the effect of a row of rooms lit by sunlight, and so enhance the appearance of the main room, even though they serve no advantage to people working in the alcoves themselves.

Mural designs are a well-known decorator's artifice, and need no discussion here except to indicate that skilled top-lighting of such a device can create the effect of a window in the roof. Here again the choice of lamp colour can provide an enhanced effect.

Large clear glass-partitioned spaces within a very large windowless enclosure are indicated wherever the industrial process does not preclude their use. There is little doubt that one of the features of a windowless building which is most disliked is the lack of a distant view. In a large building, the use of glass partitions through which a distant view can be obtained goes far to alleviate discomfort and to give a sense of orientation within the envelope. The partitions themselves create an illusion of distance. One of the reasons why the 'bürolandschaft' idea of office design has had success is that the use of acoustic partitions not only reduces noise, but creates visual illusions of distance and of interest which are absent from the large unpartitioned office.

Special distributions of light and colour used for the alleviation of enclosure-discomfort in windowless rooms is a technique at present in its first stages. Many empirical approaches to the problem have been made by designers with varying success. A systematic study being made at the Building Research Station (1966) by Plant and Collins, which is in course of publication, includes the use of colour.

Surface colour in windowless rooms requires special care in its choice, as it will, together with the lighting, determine the character of the space without any mitigating effect from the changing pattern of the outside daylight and sunlight. The 'chromatropic' properties of colour to attract and direct the attention should be employed, as Gloag has advocated, to assist people to orientate themselves in the building which has no windows to give any clue as to where exits can be expected to be found, and who therefore need such guidance especially in case of fire or other emergency.

References

British Standard 2660:1961: 'Colours for Building and Decorative Paints.'
British Standard 950:1967: 'Artificial Daylight for the Assessment of Colour. Part 1. Illuminant for Colour Matching and Colour Appraisal.'
Building Research Station (1966). Annual Report, 1965, HMSO, London.
CIE (1968). Report of Committee E–1.3.1. (Colorimetry). Proc. CIE 16th Session (Washington, 1967). Commission Internationale de l'Éclairage, Paris.

Cockram, A. H., J. B. Collins and F. J. Langdon (1969): 'A Field Study of Lamp Colour Preferences for Supplementary Lighting.' *Lighting Research and Technology*. To be published 1969.
Department of Education and Science (1967): 'Lighting in Schools.' Building Bulletin No. 33, HMSO, London.
Evans, R. M. (1964). 'Variables of Perceived Color'. *J. Opt. Soc. Amer.,* **54**, 1467–1474.
Gloag, H. L. and M. Keyte (1959). 'The Lighting of Factories.' Factory Building Studies, No. 2. HMSO, London.
Gloag, H. L. (1961): 'Colouring in Factories.' Factory Building Studies, No. 8. HMSO, London.
Gloag, H. L. (1969). 'Hue, Greyness and Weight'. Building Materials. To be published.
Hopkinson, R. G. (1952). 'The Selection of Suitable Chalkboard Colours.' RIBA J. **59**, (10), 377.
Hopkinson, R. G. (1954): 'Reflected Daylight.' *Archit. J.* **120**, (3101, 5th August), 173–177.
Hopkinson, R. G. (1957). 'Éclairage Supplementaire Permanent par Fluorescence dans les Écoles.' Journées de l'Éclairage, Lyon, Assoc. Franc. de l'Éclairage, 121–123.
Hopkinson, R. G. and J. Longmore (1959): 'The Permanent Supplementary Artificial Lighting of Interiors.' *Trans. Illum. Engng. Soc.* (*London*), **24**, (3), 121–148.
Hopkinson, R. G., J. Longmore, D. Medd and H. L. Gloag (1964): 'Integrated Daylight and Artificial Light in Interiors.' Proc. CIE 15th Session (Vienna, 1963). Paper P63.12. Commission Internationale de l'Éclairage, Paris.
Hopkinson, R. G., P. Petherbridge and J. Longmore (1966). 'Daylighting'. Heinemann, London.
Jay, P. (1968): 'Inter-relationship of the Design Criteria for Lighting Installations.' *Trans. Illum. Engng. Soc.* (*London*), **33**, 47–71.
Longmore, J. and P. Petherbridge (1961): 'Munsell Value/Surface Reflectance Relationships.' *J. Opt. Soc. Amer.*, **51**, 370–371.
Lynes, J., W. Burt, G. K. Jackson and C. Cuttle (1966): 'The Flow of Light into Buildings.' *Trans. Illum. Engng. Soc.* (*London*), **31**, 65–91.
Lynes, J., with C. Cuttle, W. B. Valentine and W. Burt (1967): 'Beyond the Working Plane.' Proc. CIE 16th Session (Washington, 1967). Paper 67.12. Commission Internationale de l'Éclairage, Paris.
Ne'eman, E. (1969) and Hopkinson, R.G.: 'Minimum Window Size.' Lighting Research and Technology. To be published.

Index

Index